abc's of Boolean Algebra

by

Allan Lytel

and

Lawrence Buckmaster

HOWARD W. SAMS & CO., INC.
THE BOBBS-MERRILL CO., INC.
INDIANAPOLIS · KANSAS CITY · NEW YORK

THIRD EDITION

FIRST PRINTING—1972

International Standard Book Number: 0-672-20867-9
Library of Congress Catalog Card Number: 77-173230

Preface

Boolean algebra is the algebra of *logic,* an abstract mathematical structure appearing in three different forms—a switching algebra, a propositional calculus, and an algebra of sets. Heady fare for a beginner, perhaps, yet you do not need a mathematical background to understand and use Boolean algebra. All you need is an orderly and inquiring mind. Boolean algebra is the basis for switching engineering, from which come modern electronic computers, automated process control, and telephone systems.

With simple block diagrams, *abc's of Boolean Algebra* shows you the relation between language and switches, the principles of logical design, and examples of the application of these principles. It tells you how to write logical expressions, how to expand and simplify them, and how to use relays and switches in simple practical circuits.

Block diagrams explain AND, OR, NAND, and NOR circuits and their symbols and show how to convert algebraic expressions into practical switching circuits. Numbering systems are discussed to show how logical circuit blocks can be combined and used to make calculations.

From simple examples of switching circuits, the text progresses through circuits of increasing complexity until finally the logical functions of entire stages are discussed. A chapter on the algebra of states and events adds the concept of time and such operators as DELAY, HAPPEN, BEFORE, AFTER, ENTERING, LEAVING, and others.

This book is for engineers, electronics technicians, students, and interested laymen. For these people the content forms a base, not only for more advanced study, but also for a deeper understanding of the complex mechanisms that do much of the work in the world of today.

Contents

CHAPTER 1

An Introduction to Logical Concepts

One of the more ironical aspects of science is that sometimes many years elapse between the construction of a theoretical system and its application in engineering and technology. For example, complex numbers, sometimes called imaginary numbers, were used only in a mathematical, abstract sense for many years before their application to alternating-current circuit design.

Another example is symbolic logic. This was a fully developed and independent field of study long before its application to digital circuit design. It was first formulated by the famous German mathematician, Gottfried Wilhelm von Leibnitz (1646-1716), with other basic contributions by Augustus De Morgan (1806-1871) and George Boole (1815-1864).

One entire field of symbolic logic is now called Boolean algebra after George Boole, an English mathematician, who in 1854 published a book entitled, *An Investigation of the Laws of Thought on Which are Founded the Mathematical Theories of Logic and Probabilities.* At the start of Chapter One, he wrote:

> The design of the following treatise is to investigate the fundamental laws of those operations of the mind by which reasoning is performed; to give expression to them in the symbolical language of a Calculus; and upon this foundation

> to establish the science of logic and construct its method; to make that method itself the basis of a general application of a mathematical doctrine of the Probabilities; and finally to collect from the various elements of truth brought to view in the course of these inquiries some probable intimations concerning the nature and the constitution of the human mind.

When it was published it was considered an abstract mathematical novelty of little consequence. In actuality, and quite probably without Boole's knowing it, this one work swept aside the tradition of nonsymbolic Aristotelian logic, and prepared the way for the precise mathematical treatment of logic with a new algebra particularly suited to handling classes and propositions. Other mathematicians, notably John Venn and Ernst Schroder, improved and expanded Boole's work.

It was first recognized as a fundamental contribution to the field of mathematics by Whitehead and Russell in their monumental *Principia Mathematica* (1910-1913). Another important work that deserves mention is the classic, *Mathematical Logic* (1928), by Hilbert and Ackermann.

The milestone in symbolic logic, indeed the turning point between the study of logic as an abstract theory and its application in modern electronics, was the publication of "A Symbolic Analysis of Relay and Switching Circuits" in 1938. Its author, Dr. Claude E. Shannon, was then a student at MIT and afterwards a well-known mathematician and engineer at Bell Telephone Laboratories. This paper, which first appeared in the AIEE *Transactions* (Vol. 57, 1938), was an abstract of Shannon's thesis for the degree of Master of Science. Comprising only ten pages, this paper was the tap root from which has come much of the work in symbolic logic, and the basis of the logical design of digital circuitry.

LOGICAL CONCEPTS

All of us have had some familiarity with logical concepts whether or not we are familiar with the special rules and vocabulary of symbolic logic. Our language, our very thinking process, uses logic in a fuzzy sort of way, and while we may not be aware of it, we have had already a good deal of experience with its content. The content of symbolic logic includes the ideas and relationships expressed by the words in the following list.

TRUTH, FALSITY
ASSERTION, DENIAL
GROUPS, INDIVIDUALS, ELEMENTS
STATEMENTS, PROPOSITIONS
CLASSES, COLLECTIONS, TYPES
CHOICE, SELECTION, COMPARISON,
 ARRANGEMENT
MERGING, SORTING, COLLATING

Unfortunately, our use of these words, as well as most words in general, is often quite inexact. Unless we know its context, a word may have several different meanings. The word *yes*, for example, can be variously interpreted. It can means, "yes, the statement is true." But it can also mean, "yes, probably," or even, "yes, you have my attention; what do you want?"

Symbolic logic assigns precise meanings to words and their associated symbols. A symbol in a logical context will have but one possible definition regardless of its context, and that definition is even stabilized and sharpened by exact relationships with other precise symbols. For example, in symbolic logic, "the" is an assertion meaning, "there exists one and only one." So, "the white house" implies the assertion, "there is one and only one white house."

WHAT IS SYMBOLIC LOGIC

Symbolic logic in a broad sense is a science that has the following characteristics:

(A) it involves non-numerical relationships
(B) it seeks precise definitions
(C) it results in inevitable and fixed consequences
(D) it uses rigidly defined and efficient symbols

While usually considered a part of mathematics, symbolic logic differs from mathematics in a number of ways. For example, symbolic logic generally does not include numerical relationships like, "plus, minus, greater than," or indefinite qualities like, "several, many, most," or even numerical quantifiers like "four," or "six hundred." These concepts are more properly a part of mathematics.

Mathematics is concerned with:

(A) numbers—such as 8, π, $2/3$, 8.59
(B) shapes—such as triangles or cubes
(C) arrangements—such as the sequence 1, 4, 9, 16, 25
(D) patterns—such as a floor plan

Symbolic logic deals with:

(A) statements—switch L is set at position B
(B) classes—such as switches, indicators, contacts
(C) relationships—switch V is on only if switch R is set at G
(D) properties—such as magnetic, fast-acting, time-delayed

Typically, mathematics deals with questions like, how much, how many, how far. Symbolic logic deals with questions such as, what is the result of this action, does this statement have any exceptions, or what is the proof for this theorem.

For example, a rule in mathematics is "the shortest distance between two points is a straight line." A rule in symbolic logic is "the denial of the denial of a statement is the statement itself."

Originally, symbolic logic was designed as a technique of semantics and instruction in the use of language; it was designed to provide an analytical and logical method of presenting ideas. For this reason, symbolic logic has long been taught in colleges and universities as a language technique.

There are several fairly well recognized divisions of symbolic logic in use today. One of the more important divisions is Boolean algebra, the algebra of AND, OR, NOT, and of conditions, statements, classes, states, and events. Boolean algebra has proven to be invaluable in designing electrical circuits using relays, electronic tubes, or transistors used in digital computers. These applications are discussed at length in subsequent chapters, but it is possible to make a brief and yet meaningful introduction to the field in terms of its use in control systems.

Suppose that there are two switches connected in series to a source of electrical energy, such as a battery. Label these two switches A and B. This may be called an AND circuit, since it is quite obvious that there will be current only if A *and* B are both closed. This seems to be a very trivial point, yet it is the very basis of symbolic logic as used in electronic switching circuits. In logical arrangements, an output from the circuit is desired only if A is true (switch A closed), and B is true (switch B closed). It is not enough that either switch alone be closed; it is important that they both be closed at the same time.

Another possibility to consider again uses two switches, but here the switches are connected in parallel and each is in series with a source of energy. This can be called an OR circuit because there will be current if A is closed *or* if B is closed. Thus, if it is desired to have an output if *either* input is present, a parallel arrangement, known as an OR circuit, is used.

Although AND and OR circuits are obvious and quite simple, amazingly complex systems can be built with them. Obviously we could have any number of switches in series, such as A *and* B *and* C *and* D *and* E. In this case, there will be an output only if all the switches are closed. In the other case, if the switches are all connected in parallel, there will be a circuit output if any one of A *or* B *or* C *or* D *or* E is closed.

Although this is quite simple and basic, almost immediately trouble appears. Consider the OR circuit, for example; in the English language there are two meanings for the word *or*. We have a non-exclusive *or;* by this we mean it is possible that either A is true or B is true or that they are both true. As an example, it is possible to say "You are going to wear a hat or a coat." Now though it is quite clear from this sentence that you can wear a hat or you can wear a coat, it is also quite possible that you can wear both. However, it is also possible to say "I am going to New York or I am going to Chicago." Here is a very different meaning. I can go to New York or I can go to Chicago, but quite obviously I cannot go to both, at least not at the same time. The second meaning of the word *or* then, is the exclusive one, meaning either statement A or statement B is true but not both statements are true.

We can now arrive at a basic definition of symbolic logic. Symbolic logic is the development of general principles of rational procedure using ideographic symbols, in a form that demonstrates the relationships of these symbols to one another. Boolean algebra is often considered to be identical with symbolic logic, though it is actually only a small part of the whole of symbolic logic, specifically the algebra of AND, OR, NOT, and of conditions, statements, and classes.

We can see why this is called an algebra by comparing the laws of Boolean algebra to those of common algebra. In order to do this, there are three terms that must be defined.

The first is the *logical connective;* in logically arranged constructions certain statements are related by these connectives. For example, one such connective is *and;* it is possible to say "The book is green, and it has two hundred pages." The first statement is "The book is green" and the second statement is "It has two hundred pages"; the connective *and* ties the two statements together. Another logical connective is *or;* we say, for example, that it will either rain or snow. Again, there are two possible states; one is rain, the other is snow, and they are connected by the word *or.*

The second term that must be understood is *symbol.* Just as in ordinary algebra, we use symbols in the vocabulary of Boolean algebra. It is possible, for example, for A to represent the statement that the book is green and B to represent the statement that the book has two hundred pages, so that we can write in this case A *and* B, which means the book is green and it has two hundred pages.

The third definition, that of a *truth value,* is more difficult to formulate. Various types of logical statements and propositions written in symbolic form can be tested for their truth value or correspondence to reality. As an example it is sufficient to say here that a truth value of 1 means that the statement is *true* and a truth value of 0 means that the statement is *false.*

With these very rudimentary definitions it is possible to compare algebra and symbolic logic (Table 1-1). Algebra has certain symbols which can be used to represent variables that are either dependent or independent, with numerical values that range over the entire number system from plus infinity to minus infinity. The numbers that are used in algebra can be of various types such as real, imaginary, complex, rational, irrational, integral, fractional, and many others. Letter symbols are used in symbolic logic to represent dependent and independent variables just as in algebra; but these variables are *statement variables* and represent complete phrases of a sentence. Their purpose is to reduce a statement from its *content,* expressed in words, to its *form,* expressed in symbols. Once a

Table 1-1. Comparison of Algebra and Symbolic Logic

Fundamental operations of Algebra.

Addition (+)

A + B = C

A	B	C
1	1	2
1	2	3
2	2	4
5	4	9

etc. in infinite variety.

Subtraction (—)

A — B = C

A	B	C
2	1	1
3	1	2
1	3	—2
9	4	5

etc. in infinite variety.

Fundamental operations of symbolic logic.

Conjunction (symbol •)

A • B = C
A and B = C
C is true only when A and B are both true simultaneously.
True = 1,
False = 0.
No other possibilities exist.

A	B	C
0	0	0
0	1	0
1	0	0
1	1	1

Disjunction (symbol +)

A + B = C
A or B = C
C is true whenever either A or B is true.
T = 1,
F = 0.

A	B	C
0	0	0
0	1	1
1	0	1
1	1	1

Negation

$A = \bar{B}$
A is true if B is false; A is false if B is true; A is the negation of B.

A	B
0	1
1	0

chain of reasoning is reduced to its purely formal outline, the validity of any conclusion may be determined by the truth or falsity of the component statements and the truth or falsity of the connectives between the statements. Since a statement variable can have only one of two values, true or false, and since any statement connectives can only be either true or false, they may be considered *binary*, or two-valued. The system considered in its entirety can be regarded as a *binary system*. Thus it is possible to represent these variables as a series of pairs; for example, one pair is 1 and 0, another pair is true and false, another pair is pulse and no pulse, another pair is plus voltage and minus voltage. In every case we are dealing with binary numbers in which the variable can assume only one of two possible values.

FUNDAMENTAL OPERATIONS

Referring again to Table 1-1, if A is 0 and B is 0, then C is 0. If A is 0 and B is 1, C is still 0. If A is 1 and B is 0, C remains 0, but if A is 1 and B is 1, then C is equal to 1. Thus, for *conjunction,* A AND B equals C is true if and only if A and B are simultaneously true. This is written as $A \bullet B = C$.

Disjunction, which has the symbol +, is the equivalent of the word *or,* so that we can say A or B equals C. As shown in the truth table, C is 1 (C is true) if A is 1 (true) or B is 1 (true) or if both A and B are 1 (true).

The third fundamental operation is *negation;* very simply, it is stated that A is true if B is false, and, conversely, A is false if B is true. Therefore A is the negation of B as shown in the truth table.

There are other types of logical operations that are useful to the logician. But these may always be derived from the three above. These three are easily utilized by electronics; consequently, digital logic for switching circuits is generally concerned only with the operations of conjunction, disjunction, and negation.

The Binary Variable

The binary variable assumes two and only two values. It corresponds directly to the "bit" of information theory. The term *bit* means the binary digit. The two values of a binary variable are commonly represented as true, false; one, zero; plus voltage, minus voltage; pulse, absence of pulse; open relay contacts, closed relay contacts; etc. A finite number of binary variables when taken together must yield a finite number of possible combinations. Thus, if a variable is represented by the position of a toggle switch, two switches (representing two variables) can result in only four possible combinations, as shown in Table 1-2. Three switches would give eight possible combinations, four would give sixteen. The number of possible combinations can be seen to be 2^N where N is the number of switches or binary variables.

Consider the three binary variables represented by the three

Table 1-2. Combination of Two Binary Variables

A	B	
1	1	1 = ON
0	1	0 = OFF
1	0	
0	0	

switches, A, B, and C. There are eight possible configurations of ON and OFF that could be set up on A, B, and C. These are shown in Table 1-3. Some arbitrary action could be predicated solely on the occurrence of any one of the eight possible combinations of A, B, and C. The predicated action could also be based on the occurrence of more than one combination; that is, it could be initiated by either combination 1 or combination 5; or by any of combinations 3 or 4 or 7; or any combination except 8; etc. These combinations of combinations represent functions. There are a finite number of possible functions of N variables. The reader should satisfy himself that in general the number of combinations is 2^N, where N is the number of variables, and that the number of functions is 2^{2N}.

There could be any of eight occurrences in a scheme where each unique action corresponds to a single combination as 010, 110, or 001. Note that there are N variables, hence 2^N combinations. Since:

$$N = 3$$
$$\text{then } 2^N = 8$$

Also, the number of possible functions or combinations of combinations is given by:

$$n = 2^{2N}$$

Table 1-3. Combination of Three Binary Variables

Combination	A	B	C
1	0	0	0
2	0	0	1
3	0	1	0
4	0	1	1
5	1	0	0
6	1	0	1
7	1	1	0
8	1	1	1

where n is the number of possible functions.
For N = 3, we have:

$$n = 2^8$$
$$= 256 \text{ possible functions.}$$

LOGICAL CONNECTIVES APPLICATIONS

The rules of binary arithmetic provide for a complete system of numbers that may be used for counting and computation. There are several advantages to this two-valued system, as compared to systems having other numbering bases. The rules are direct and simple; for example, there are no tables for multiplication. For use in computers, binary numbers have the advantage of requiring only two states of the counters (vacuum tubes or transistors). The 1 and 0 of the binary system may be used to represent the ON and OFF states, respectively.

The manipulation of numbers can be treated by the use of logical connectives; i.e., OR or AND. As mentioned earlier, in the terms of logic, *or* has two meanings. If A is one event and B is another event, then A or B can mean either one event A, or, if not A, then the event B. But this is *not* the only meaning of the word or. If event A occurs, and at the same time event B occurs, this satisfies the condition *either* A *or* B. Hence the two meanings: either A or B *but not both,* and either A or B *or both.* Thus if A and B are both statements, we can form several compound statements.

A AND B = C

We say that C is true if (and only if) A is true and B is true. C is false if A is false, if B is false, or if both are false.

A OR B = C

We say that C is true if A is true or if B is true or (in this case) if both are true. Note that for the OR connective only one statement must be true; for the AND connective all statements must be true.

Since nearly all storage devices and circuit techniques that are used for digital systems are binary in nature, the simplicity of arithmetic operations in the binary system provides theoretical advantages and circuit simplifications that surpass those of other numbering systems. The two-state nature of binary switching devices permits the use of conventional logic; the names of the states are purely arbitrary; for the present discussion the two states will be denoted "1" and "0." State 1 is defined as the transmission of information, and state 0 as the absence of transmission. Thus a conventional switch with a pair of normally open and a pair of normally closed contacts, called A and $\overline{A}$ respectively, would be described by:

$$A = 1 \qquad \overline{A} = 0 \quad \text{when operated,}$$

and by:

$$A = 0 \qquad \overline{A} = 1 \quad \text{when not operated.}$$

Negation ($\overline{A}$) has been introduced in these expressions to define a *not on* condition (symbolized by the use of the bar). If two switches are such that one is in the transmitting state while the other is not transmitting, each device is said to be the negation of the other. A permanently closed circuit is, therefore, the negation of a permanently open circuit:

$$\overline{1} = 0 \quad \text{and} \quad \overline{0} = 1$$

Since a double negation is equivalent to an assertion (two wrongs make a right, so to speak), it follows that:

$$\overline{\overline{A}} = A$$

Table 1-4 shows the basic Boolean relations; these will be examined later and are given here for reference.

The algebraic notation for the operations is: AND is represented by the dot, i.e., A AND B is written A • B; OR is represented by the plus, i.e., A OR B is written A + B; NOT is repre-

Table 1-4. Basic Logic Rules

$1 + 0 = 1$	$X + 0 = X$	$X + \overline{X} = 1$
$1 + 1 = 1$	$X + 1 = 1$	$X + X = X$
$1 \cdot 1 = 1$	$X \cdot 0 = 0$	$X \cdot X = X$
$1 \cdot 0 = 0$	$X \cdot 1 = X$	$X \cdot \overline{X} = 0$
$0 \cdot 0 = 0$		

De Morgan's Theorem

$$\overline{(X \cdot Y)} = \overline{X} + \overline{Y};$$
$$\overline{X + Y} = \overline{X} \cdot \overline{Y}$$

Absorption Property

$$X = (X + Y) =$$
$$X + X \cdot Y = X$$

Useful Identities

$$X + \overline{X} \cdot Y = X + Y$$
$$X \cdot (\overline{X} + Y) = X \cdot Y$$
$$(X + Y) \cdot (X + Z) \cdot (X + Z) =$$
$$(X + Y) \cdot (X + Z) =$$
$$X \cdot Y + \overline{X} \cdot Z + Y \cdot Z = X \cdot Y + \overline{X} \cdot Z$$

sented by the bar, i.e., NOT A is written $\overline{A}$. The axioms of Boolean algebra are similar to those of conventional algebra:

$$\left.\begin{aligned} X + Y &= Y + X \\ X \cdot Y &= Y \cdot X \end{aligned}\right\} \text{Commutative Laws}$$

$$\left.\begin{aligned} X + (Y + Z) &= (X + Y) + Z \\ X \cdot (Y \cdot Z) &= (X \cdot Y) \cdot Z \end{aligned}\right\} \text{Associative Laws}$$

$$\left.\begin{aligned} X \cdot (Y + Z) &= (X \cdot Y) + (X \cdot Z) \\ X + Y \cdot Z &= (X + Y) \cdot (X + Z) \end{aligned}\right\} \text{Distributive Laws}$$

The relationship between the OR and the AND may be seen from the equation known as *De Morgan's theorem*. This theorem states basically: the negation of the sum of two classes is equal to the product of their negations, and the negation of their product equals the sum of their negations. In other words, to negate an expression such as $(A + B)$, negate each variable and change each AND to OR, and each OR to AND.

Thus: if we take $(A + B)$ and negate it, we get $(\overline{A + B})$. This equals $\overline{A} \cdot \overline{B}$.

Also: $A \cdot B$ negated may be written as $(\overline{A \cdot B}) = (\overline{A} + \overline{B})$.

ELECTRONIC CIRCUITS

Electronic circuits may be readily designed to meet the Boolean input and output conditions. The three basic blocks are AND, OR, and NOT, as in Fig. 1-1. In these circuits, the state of

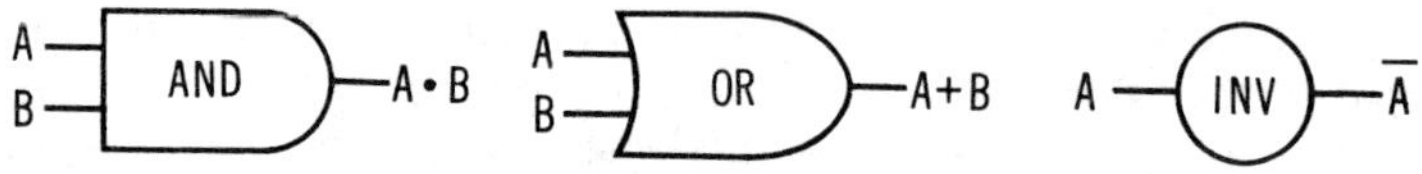

Fig. 1-1. Basic building blocks.

transmission of information (1 state) is defined by a particular voltage level and the state of nontransmission (0 state) as a second level. Typically, a ground condition represents the 1 state, and a negative voltage represents the 0 state.

A useful feature of logic circuits is variation by inversion. As De Morgan's law implies, inversion, or negation, is also applied to *operations*. As a result, the negative, or inverse, of AND is equivalent to OR. Therefore, the circuit that produces an AND for the logic convention specified before will produce an OR for inverse logic. A common variation is the use of NOT AND and NOT OR (NAND and NOR) circuit. These variations provide a powerful design tool.

CHAPTER 2

Language and Electronics

Symbolic logic may be considered as the manipulation of various statements by means of symbols, assuming that a statement is a written or a verbal assertion. An assertion is considered as a simple statement or declaration such as, "I will take a walk," or "I will read a book," or "I will go outside." It is also possible to combine two or more assertions in a sentence such as, "I will read a book if the library is open," or "I will go outside if it is not raining," or "I will go to sleep if it is after twelve o'clock." The differences here between the simple declaration and the compound declaration are simply this: a simple assertion is a simple statement although combinations of two or more simple statements form a compound statement. Note that simple statements tied together by connectives make compound statements. This chapter is concerned with such statements, both simple and compound, as well as their logical order, sequence use, and meaning, and their relationship to modern electronic switching devices such as transistors, "flip-flops," etc.

These statements are given values such as true or false; the symbolism will be T for true and F for false. Consider as an example the assertion "I will go outside." In the context of this chapter, if you go out after making such a statement, the statement is true. If you do not go out after making such a statement, the statement is false. To make the manipulation more direct and less cumbersome we will use letters such as A, B, C,

Table 2-1. Symbols of the Propositional Logic

Symbol	Name	Means
•	Conjunction	A and B
∨	Disjunction (inclusive)	Either A or B or both
⊻	Disjunction (exclusive)	Either A or B but not both
~	Negation	Not A
⇆	Equivalence (biconditional)	A if, and only if B

or D to stand for individual simple statements. Thus these letters are variables, and in the context of this chapter the variables can assume only one of two values; they can be either true or false.

In order to use statements in their compound form it is necessary to provide links or connectives. A list of these connectives is given in Table 2-1.

The first thing to remember about logical connectives is that the truth value of *any* connective is *defined* by the truth values of the propositions it connects. For example, the truth value of AND may be defined by the truth table of the conjunctive proposition A • B as shown in Table 2-2. In the first case, if A is true and B is true, the compound statement A AND B must be true. However, if, as in the second step, A is true and B is false, the conjunction of A AND B must necessarily be false. In the third example, if A is false and B is true, their conjunction in this case must be false. In the fourth case, where A is false and B is false, the conjunction of A AND B must also be false. In summary, when we have two simple statements connected by the conjunction, the compound statement A AND B is true if, and only if, A is true and B is true. There are no other possibilities; the table exhausts all the possible truth values of the propositions.

Table 2-2. Truth Table of A • B

A	B	A • B
T	T	T
T	F	F
F	T	F
F	F	F

Table 2-3. Truth Table of A ∨ B

A	B	A ∨ B
T	T	T
T	F	T
F	T	T
F	F	F

The second connective is the OR, which is the disjunction of A and B. If we have two simple statements A and B, the compound statement A OR B is true when either of the two individual simple statements is true. For example, if A is true, then A OR B is true; if B is true, then A OR B is true. However, if A is false and B is false, then the compound statement A OR B must necessarily be false. There is a problem with the case shown in Table 2-3 where A is true and B is true. The question is simply this: does the statement A OR B mean either A OR B, or does it mean either A OR B or both? If we consider that it means the latter, then A OR B is true if A is true and if B is true. In technical terms this truth table reflects the *inclusive* disjunction of A and B.

In order to differentiate between the two possible meanings of the word OR, it is necessary to define the *exclusive* disjunction of A and B as meaning either A or B but not both. A truth table for this is shown in Table 2-4. By definition, the only two cases that are true are the second and third shown in the truth table. In case 2, A is true and B is false; in case 3, A is false and B is true. Both of these compound statements are then true. In the fourth case, if both A and B are false, the compound statement is false. In the first case, by definition, if both are true the compound statement is false, since the exclusive disjunction of A and B is necessarily false if both are true.

Another basic connective is *negation*, which means simply that NOT A is the negation of A, and NOT B is the negation of B. This means, of course, that if A is true, the negation of A is

Table 2-4. Truth Table of A $\underline{\vee}$ B

A	B	A $\underline{\vee}$ B
T	T	F
T	F	T
F	T	T
F	F	F

Table 2-5. Truth Table of A ∨ ∼ B

Consider A ∨ ∼ B (A or not B)		
A	B	A ∨ ∼ B
T	T	T ∨ F is T
T	F	T ∨ T is T
F	T	F ∨ F is F
F	F	F ∨ T is T

false. If A is false, the negation of A is true. Although the fundamental idea of negation seems simple it is very significant for compound statements. For example, Table 2-5 shows the statement A OR NOT B. In the truth table shown, there are again four possible cases. Consider the first, where A is true and B is true. Since B is true, its negation is false, and the statement becomes the equivalent of a compound statement that is true OR false; hence the statement is true. In the second case, when A is true and B is false, the negation of B is of course true; hence the statement becomes a compound statement that is true OR true, which of course is true. In the third case, A is false and B is true; the negation of B is then false so that this compound statement is false. In the last case, A is false and B is false; the negation of B is then true so that the compound statement is false OR true, which is true. This is an example of the inclusive disjunction of A OR NOT B. For convenience in printing, negation is sometimes shown by a dash over that which is negated. NOT A can be shown by ∼A or by $\overline{A}$.

A more complex compound statement (Table 2-6) is NOT [(A

Table 2-6. Truth Table of Compound Statement

∼[(A ∧ B) ∨ (A ∧ ∼ B)]

not [(A and B) or (A and not B)] = C

not (A and B) and not (A and not B) = C

(not A or not B) and (not A or B) = C

A	B	(A ∧ B)	(A ∨ B)	(A ∧ ∼ B)	(A ∧ B) ∨ (A ∧ ∼ B)	C
1	2	3	4	5	6	
T	T	T	T	F	T	F
T	F	F	T	T	T	F
F	T	F	T	F	F	T
F	F	F	F	F	F	T

AND B) OR (A AND NOT B)]. For statements like this, in the preparation of the truth table, it is a simplification to have several tabular entries. For example, the first two columns represent all possible cases for A and for B. The third column represents the truth value of A AND B; the fifth column represents the truth value of the second parenthetical expression which is (A AND NOT B). (The fourth column is included merely to enable the reader to derive the fifth and sixth columns.) The sixth column represents the entire statement (except for the initial negation) which is [(A AND B) OR (A AND NOT B)]. The last column, written as C for convenience, represents the entire compound statement. It gives the truth value of the entire statement.

Consider the first case, where A is true and B is true. Here obviously A AND B is a true statement, though A AND NOT B is a false statement. Hence, the disjunction in the sixth column, which is the disjunction of a true statement and a false statement, is true. The entire compound statement in brackets is negated, hence C is false. In the same manner with the second example, where A is true and B is false, it follows that A AND B is false, though A AND NOT B is true, so that [(A AND B) OR (A AND NOT B)] is true, but again the entire statement is negated; hence, C is false. In the third case, where A is false and B is true, A AND B is false, and it follows that (A AND NOT B) is false, so that [(A AND B) OR (A AND NOT B)] is also false; the final negation makes the entire statement true. In the last case, where A is false and B is false, (A AND B) is false; (A AND NOT B) is false; [(A AND B) or (A AND NOT B)] is false; the negation makes statement C true.

Truth tables of this type require careful thought. Another example is shown in Table 2-7, where NOT [(A OR B) AND (NOT

Table 2-7. Truth Table of C = NOT [(A OR B) AND (NOT A AND B)]

$\sim [(A \vee B) \wedge (\sim A \wedge B)]$

A	B	$(A \vee B)$	$(\sim A \wedge B)$	$[(A \vee B) \wedge (\sim A \wedge B)]$	C
T	T	T	F	F	T
T	F	T	F	F	T
F	T	T	T	T	F
F	F	F	F	F	T

Table 2-8. Conditional Connectives

Conditional	Biconditional
$\supset$ if A then B	$\leftrightarrows$ A if, and only if, B

A AND B)] is represented by C. In the first line, where A is true and B is true, it follows that A OR B is true, and (NOT A AND B) is false; thus [(A OR B) AND (NOT A AND B)] is false, so that the negation makes statement C true. In the second case, where A is true and B is false, A OR B is true, though (NOT A AND B) is false, so that the statement [(A OR B) AND (NOT A AND B)] is false; hence C is true. In the third case, where A is false and B is true, (A OR B) is true; (NOT A AND B) is true, and [(A OR B) AND (NOT A AND B)] is true. Therefore C is false. In the fourth case, where A is false and B is false, (A OR B) is false, (NOT A AND B) is false. Therefore [(A OR B) AND (NOT A AND B)] is false, and so C is true.

There are two additional connectives (Table 2-8) which are required. The first of these is the conditional "IF A THEN B," while the second is the biconditional "A IF, AND ONLY IF, B." As shown in the first case (conditional connective) in Table 2-9, A is true and B is true. From this it follows that IF A THEN B must be true. We can also see that in the second case, where A is true and B is false, the expression is false for the statement IF A THEN B. The reason why the first case is true and the second false lies in the definition of implication.

However, there is a problem where A is false and B is true as in case 3, or if both are false as in case 4. In case 2 the statement is false, since a true proposition cannot imply a false one. In cases 3 and 4, the statement is true, since a false proposition implies *any* proposition, true or false.

Remember, the truth value of the connective is completely determined by the truth values of the propositions it connects. In-

Table 2-9. Truth Table of A $\supset$ B

A $\supset$ B, if A then B

A	B	A $\supset$ B
T	T	T
T	F	F
F	T	T
F	F	T

Table 2-10. Truth Table of A ⇆ B

A ⇆ B

A if, and only if, B

A	B	A ⇆ B
T	T	T
T	F	F
F	T	F
F	F	T

deed, it is only in the truth table that a definition for implication is found. There is no necessary relationship between the logical proposition A ⊃ B and the real world, although our feelings may tell us otherwise. The realization of this fact is an essential step in the understanding of symbolic logic.

The biconditional connective is shown in the truth table in Table 2-10. This is "A *if, and only if,* B." By definition, this compound statement is true only under the circumstances of case 1 or 4, where A and B have the same truth value. In case 1, A is true and B is true; in case 4, A is false and B is false. In both these cases the compound statement is true. When A and B differ as in cases 2 and 3, the compound statement is false. In 2, where A is true and B is false, the bioconditional statement is false. In 3, where A is false and B is true, the biconditional statement is false.

ELECTRONIC SWITCHES

A truth table is a list of all possible values for the independent variables in a logical proposition. These values can also be shown in terms of their electronic implementation. For example, consider Fig. 2-1; a simple OR arrangement. Again we define 1 as representing a closed switch, 0 as an open switch, and OR as the logical connective in the proposition A + B. In its circuit implementation we can consider this proposition as representing two switches in parallel. The circuit illustrations and the accompanying truth table indicate all the possible combinations for this logical proposition as well as its electronic analogy.

If both switches are open, which is represented in the truth table for A equals 0 and B equals 0, then there will be no current in the circuit. This is represented by the case in which C

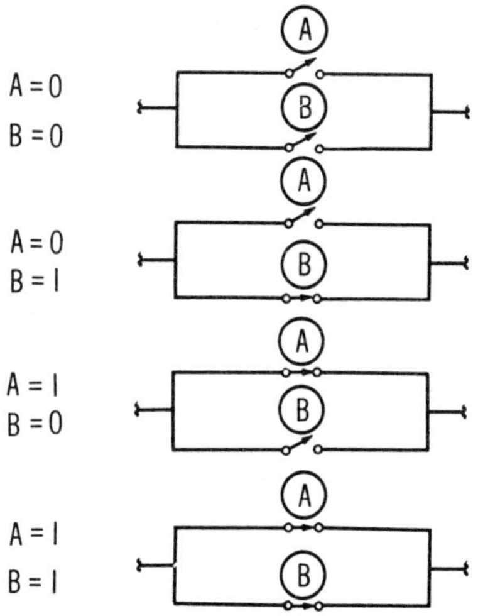

C = A+B

A	B	C
0	0	0
0	1	1
1	0	1
1	1	1

Fig. 2-1. OR circuit using switches.

equals 0 in the truth table. If A is open and B is closed, then there will be current in the circuit; this is represented by the second case in the truth table, where A equals 0, B equals 1, and C equals 1. In the third circuit, since A is closed and B is open, this will result in current, so once again C equals 1. In the fourth case, where A is closed $(A = 1)$ and B is closed $(B = 1)$, current will flow, and C is equal to 1. Thus we can see that current will flow $(C = 1)$ when switch A is closed $(A = 1)$ OR switch B is closed OR when both switches are closed $(A + B = 1)$.

Fig. 2-2 shows the AND circuit. In terms of electronics this can be two switches in series. Once again these two switches are A and B, and the logical statement is "C is equal to A AND B" $(C = A \bullet B)$. Again there are four simple circuits and a

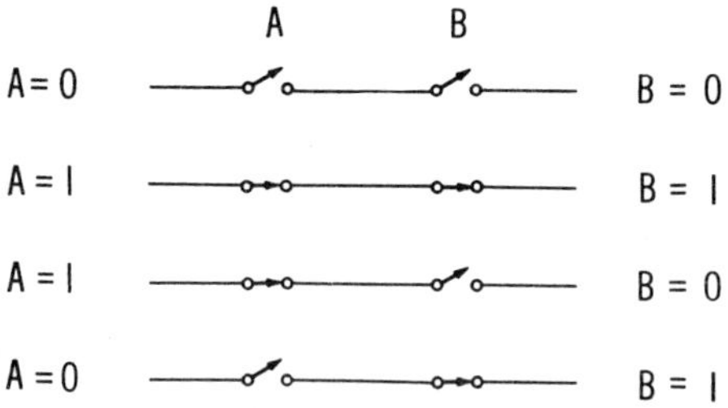

C = A·B

A	B	C
0	0	0
1	1	1
1	0	0
0	1	0

Fig. 2-2. AND circuit using switches.

truth table. As in the previous example, A = 0 represents an open switch A, and A = 1 a closed switch. When switch A is open (A = 0) and switch B is open (B = 0), no current will flow and C, an open circuit, equals 0. When switch A and switch B are closed (A = 1 and B = 1), current will flow, and C = 1. However, when either of the switches is open (either A = 0 or B = 0), the circuit is not complete, and C = 0. Therefore, we can see that in this electronic analog of A • B, both A *and* B must be closed (A = 1, B = 1) to obtain a complete circuit (C = 1).

CIRCUITS AND EQUATIONS

We shall now consider a relationship among three different forms of logic. These are the logical equation, the Venn diagram, and the electronic switching circuit. Fig. 2-3 shows a logical equation where C is a function of both A and B.

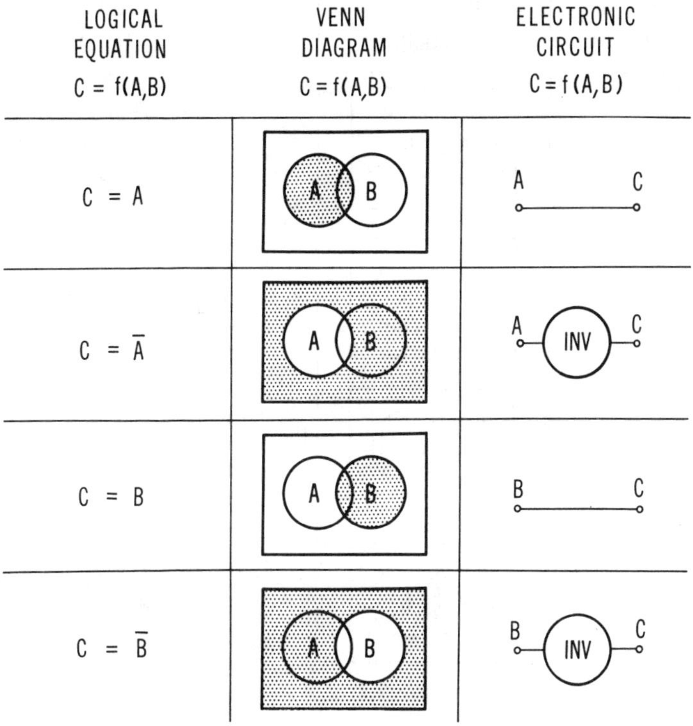

Fig. 2-3. Three different forms of logic.

The first equation is $C = A$. In the Venn diagram the darker area represents A. Implementation of this circuit is a simple straight wire since everything that happens to A also happens to C. The second equation is $C = \overline{A}$ (C is equal to NOT A). In the Venn diagram everything but the area within the circle A is darkened to represent NOT A. The electronic implementation of "C equals NOT A" requires an inverter so that A will always be the negation of C. In terms of the binary variables, if A is 1, NOT A is 0, hence C is 0. If A is 0, NOT A is 1, so that C is 1. The third equation, $C = B$, is similar to $C = A$, while the fourth, $C = \overline{B}$ (C is equal to NOT B), is similar to the equation $C = \overline{A}$.

Fig. 2-4 shows a group of equations using the logical connective AND. The first one is $C = A \bullet B$ (C is equal to A AND B). In the Venn diagram the area that is in common to both A and B is the darkened area in which the two circles A and B

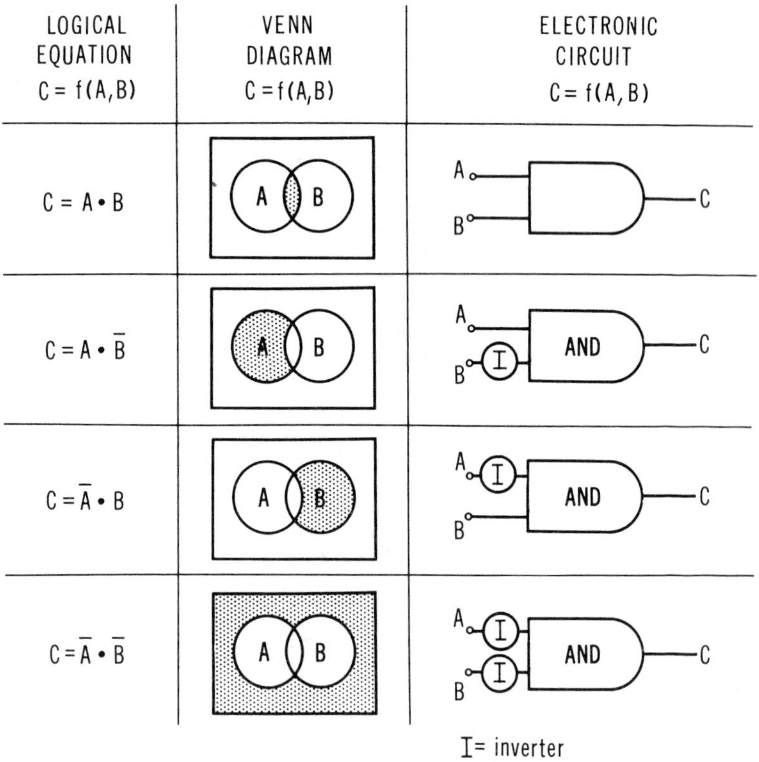

Fig. 2-4. Three forms of logic using AND.

overlap. This is C, the only area in the diagram that is common to both A and B. The electronic circuit for this condition would be the very simple AND gate in which the two inputs A AND B must both be present simultaneously in order to get an output at C.

The second logical equation is $C = A \cdot \overline{B}$ (C is equal to A AND NOT B). Note in the Venn diagram that the entire circle A is included except that portion of A which overlaps the circle B. The logical block diagram or electronic implementation is the same AND gate as above except that B is inverted. The inputs to the AND gate are A AND B, but since B is inverted the output C is equal to A AND NOT B.

The third equation is $C = \overline{A} \cdot B$ (C is equal to NOT A AND B). Note in the Venn diagram the area described in the equation is the circle B less that part of B which overlaps the circle A. In the electronic implementation or block diagram, A is inverted to produce the input NOT A, and B is applied directly to the AND gate to develop NOT A AND B.

In the fourth case, the Venn diagram shows that the area C is the entire rectangle except for the circles including the areas A and B, and the area in which they overlap. This circuit is implemented electronically by inverting A and inverting B and putting both of them through the gate to produce NOT A AND NOT B ($\overline{A} \cdot \overline{B}$).

In a similar manner, various types of OR circuits are shown in Fig. 2-5. The implementation for this logical equation is through various forms of the OR gate. In the first case $C = A + B$ (C is equal to A OR B), the Venn diagram for this shows quite clearly that this is the same as the inverse of $C = \overline{A} \cdot \overline{B}$ (C is equal to NOT A AND NOT B). The electronic block diagram for this is a simple OR circuit with A and B as the inputs.

The second logical equation is $C = \overline{A} + B$ (C is equal to NOT A OR B). As shown in the Venn diagram this is the inverse of $C = A \cdot \overline{B}$ (C is equal to A AND NOT B). Implementation is obtained by inverting A input.

To obtain $C = A + \overline{B}$ (C is equal to A OR NOT B) we invert the B input. As shown in the Venn diagram this is the inverse of $C = \overline{A} \cdot B$ (C is equal to NOT A AND B).

The last logical equation shows $C = \overline{A} + \overline{B}$ (C is equal to NOT A OR NOT B). This can be obtained electronically by inverting

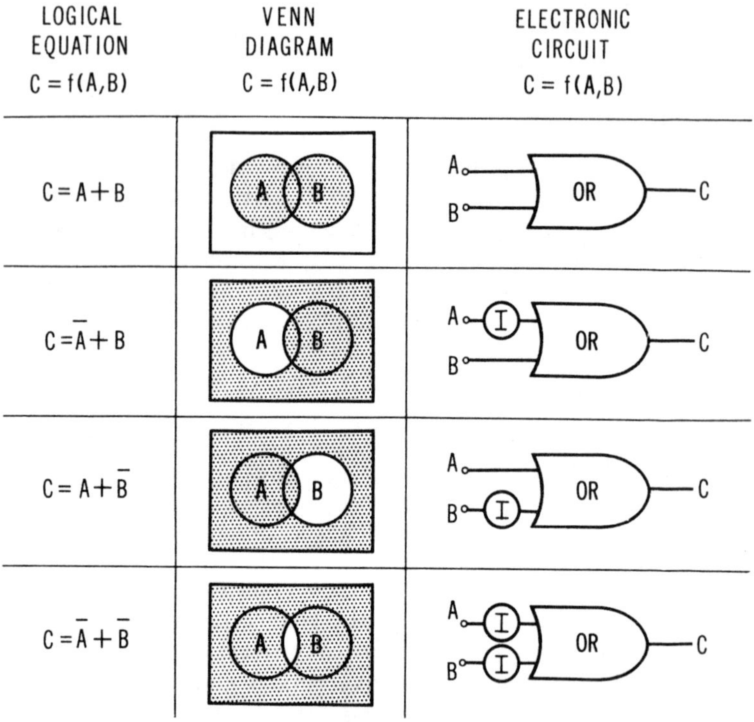

Fig. 2-5. Three forms of logic using OR.

both A and B inputs into the OR gate. This is the inverse of $C = A \cdot B$ (C is equal to A AND B).

Fig. 2-6 shows some other possible logical arrangements. The first equation is [(A AND NOT B) OR (NOT A AND B)]. As shown in the Venn diagram this includes the areas of circle A and circle B but not the area that is common to both. This is implemented in the block diagram by having one AND circuit to produce A AND $\overline{B}$ (A AND NOT B), another AND circuit to produce $\overline{A} \cdot B$ (NOT A AND B), and an OR gate to produce the final results of [(A AND NOT B) OR (NOT A AND B)].

The second equation is [(C is equal to A AND B) OR (NOT A AND NOT B)]. In the Venn diagram this includes the whole area less the area A and less the area B, but including the area which is in common to both. This is implemented electronically by using two AND gates: the first produces A AND B, the second produces NOT A AND NOT B, while these are both used as input

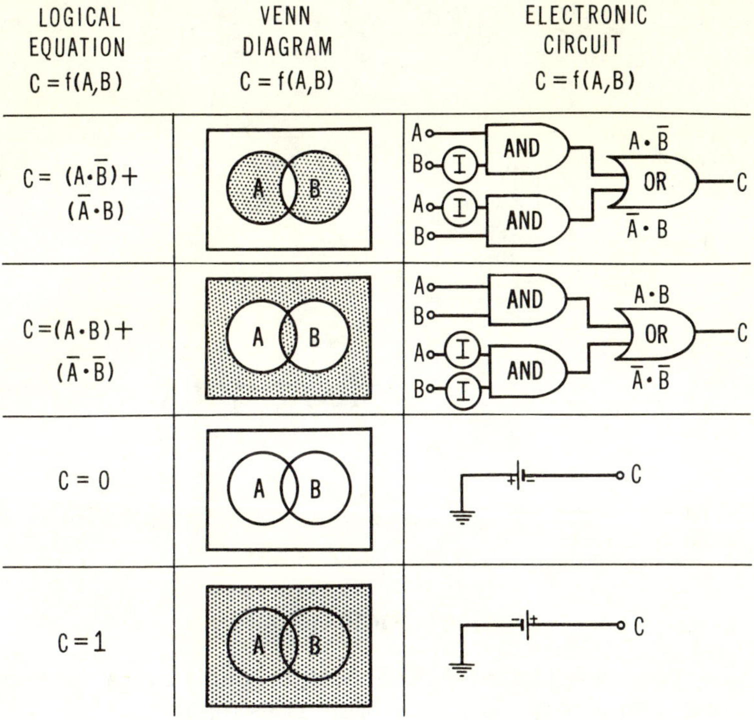

Fig. 2-6. Three forms of logic using connectives.

into an OR gate whose output is [(A AND B) OR (NOT A AND NOT B)].

The third case is C = 0 (C is equal to 0) which means that the set C is empty, or C has no occupants. The last case is C = 1 (C is equal to 1) that shows that the set C contains all possible occupants.

CHAPTER 3

Logical Circuits

Ordinary switches provide a simple implementation for basic logical circuits.

RULES OF LOGICAL DESIGN

In order to apply the logical concepts to circuits, some symbolism is necessary. This symbolism is as follows.

Symbol	Logic	Relay and Contact	Meaning
1	True	Closed	The statement is true, the circuit is closed.
0	False	Open	The statement is false, the circuit is open.
•	Series	A and B	A is in series with B.
+	Parallel	A or B	A is in parallel with B.

Before any circuit analysis can be made, some of the more fundamental relations should be established. A set of contacts is denoted by a letter such as A, B, or C. In some cases, to express an unknown, the letters may be X, Y, or Z. If 0 means open and 1 means closed, the following relations apply:

$0 + 0 = 0$ An open in parallel with an open is open.
$0 + 1 = 1$ An open in parallel with a closed is closed.
$1 + 1 = 1$ A closed in parallel with a closed is closed.

And in the same manner for series:

$0 \bullet 0 = 0$ An open in series with an open is open.
$0 \bullet 1 = 0$ An open in series with a closed is open.
$1 \bullet 1 = 1$ A closed in series with a closed is closed.

Since contacts are either open or closed, only two symbols are necessary to denote the states of the contacts. We use 1 for closed contacts and 0 for open contacts. Suppose two contacts are in series, as in Fig. 3-1, where X denotes the state of the first and Y the state of the second. Contacts in parallel are shown in Fig. 3-2. The possible combinations in this arrangement are:

Fig. 3-1. X AND Y network.

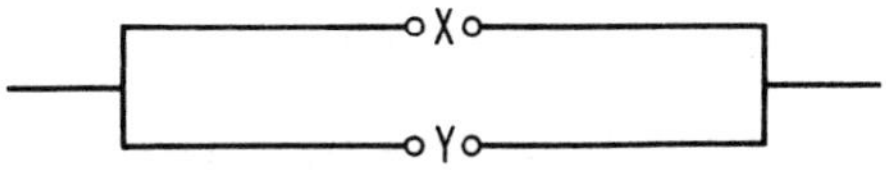

Fig. 3-2. X OR Y network.

X	Y	Network X + Y
0	0	0
0	1	1
1	0	1
1	1	1

To prove $1 + X = 1$, we set $X = 0$, getting $1 + 0 = 1$; setting $X = 1$, we have $1 + 1 = 1$. Since these equations are valid, the rule is proved. As in Fig. 3-3:

$$0 + X = X$$
$$X + X = X$$

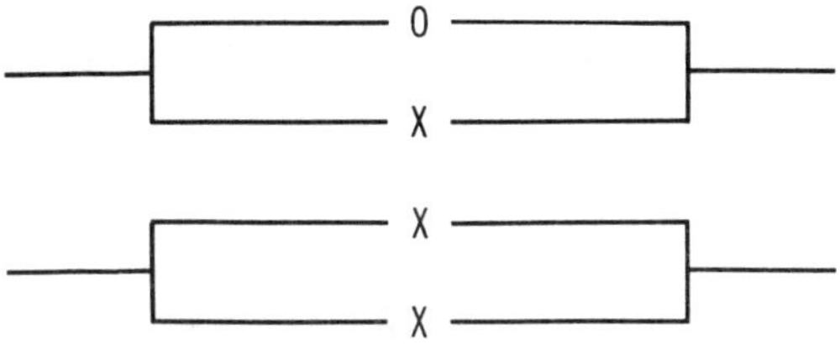

Fig. 3-3. Implementation of 0 + X and X + X with switches.

In addition to these rules, a representation of normally closed contacts as well as those normally open is required. If X denotes the normally open contacts of a relay, then $\overline{X}$ will denote normally closed contacts. We may substitute 0 as a value of X, and 1 as a value of $\overline{X}$. From this, as in Fig. 3-4:

$$\overline{0} = 1$$
$$\overline{1} = 0$$

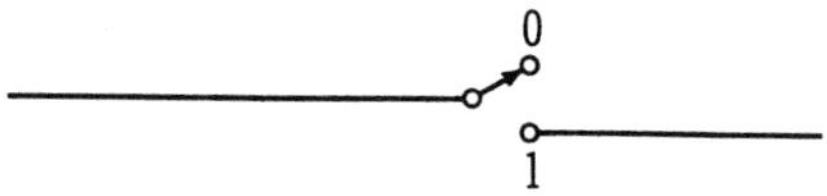

Fig. 3-4. Implementation of binary variable.

For De Morgan's law,

$$(\overline{X + Y}) = \overline{X} \bullet \overline{Y},$$

there are two values for each of the two variables, and there are four possible combinations of them. If X = 1, Y = 0, then

$$(\overline{X + Y}) = (\overline{1 + 0}) = \overline{1} = 0$$

and

$$\overline{X} \bullet \overline{Y} = \overline{1} \bullet \overline{0} = 0 \bullet 1 = 0$$

Hence, for X = 1 and Y = 0, we have $(\overline{X} \bullet \overline{Y}) = \overline{X + Y}$. The three remaining cases are proved just as easily.

The two distributive laws are represented in Figs. 3-5 and 3-6. They are:

$$X \bullet (Y + Z) = X \bullet Y + X \bullet Z$$
$$X + (Y \bullet Z) = (X + Y) \bullet (X + Z)$$

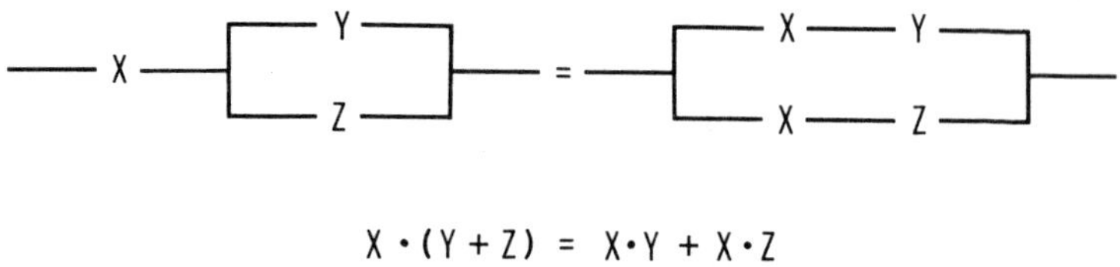

Fig. 3-5. Implementation of distributive law—first case.

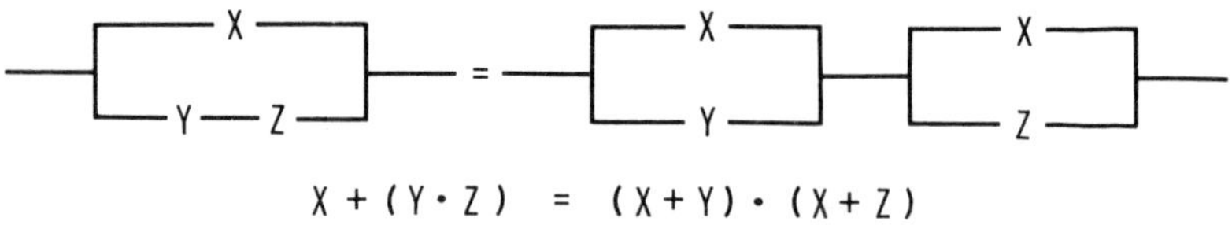

Fig. 3-6. Implementation of distributive law—second case.

We also need the associative and commutative laws:

$$X + Y = Y + X$$
$$X \bullet Y = Y \bullet X$$
$$X + (Y + Z) = (X + Y) + Z$$
$$X \bullet (Y \bullet Z) = (X \bullet Y) \bullet Z$$

Consider Fig. 3-7, where the circuit is represented by $X \bullet (X + Y)$. If X is closed $(X = 1)$, the circuit is closed; if X is open

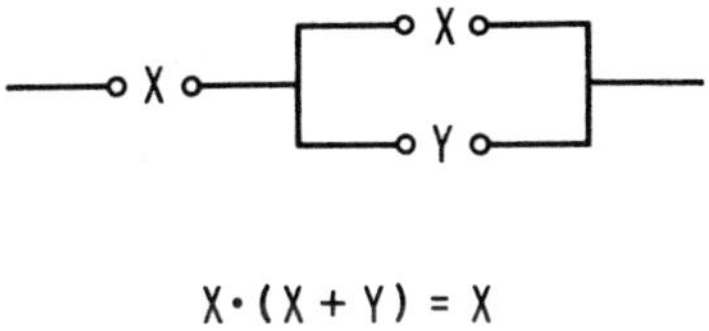

Fig. 3-7. Implementation of X • (X + Y).

$(X = 0)$, the circuit is open. Both of these statements are also true for either value of Y. A truth table is:

X	**Y**	**Network X • (X + Y)**
1	0	1
0	1	0
1	1	1
0	0	0

Thus, only if $X = 1$ is the network closed, and $X \bullet (X + Y) = X$. This may be extended to the network,

$$X \bullet (X + Y + Z) \bullet (X + Y) \bullet (Y + Z)$$

as in Fig. 3-8A. Since $X \bullet (X + Y + Z)$ reduces to X, and $X \cdot (X + Y)$ reduces to X, then

$$X \bullet (X + Y + Z) \bullet (X + Y) \bullet (Y + Z)$$

becomes $X \bullet (Y + Z)$ as shown in Fig. 3-8B.

The utility of even these few relationships may be seen in Fig. 3-9, where the relay circuit is drawn and is expressed by

$$[X + Y \bullet (X + Z)] \bullet (Y + W) \bullet Z$$

Then the circuit may be evaluated for a given set of conditions. If $W = 0$, $X = 0$, $Y = 1$, and $Z = 1$, these values may be substituted directly in the expression

$$[X + Y \bullet (X + Z)] \bullet (Y + W) \bullet Z$$

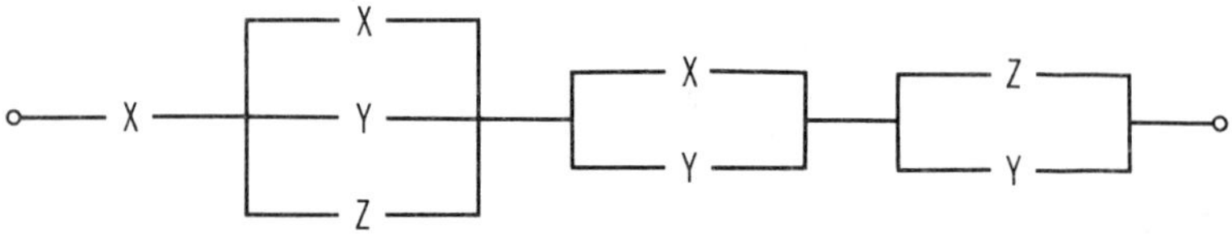

$$X \cdot (X+Y) \cdot (X+Y) \cdot (Z+Y)$$

(A) Original circuit.

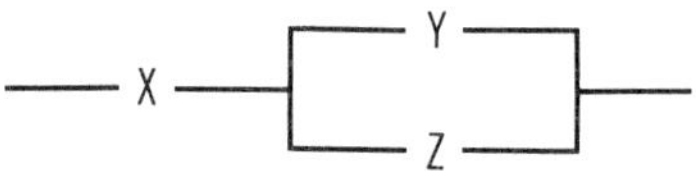

(B) Simplified circuit.

$$X \cdot (Y+Z)$$

Fig. 3-8. Implementation of commutative principles.

However, we may rewrite this expression as

$$[X + (X \cdot Y + Y \cdot Z)] \cdot (Y \cdot Z + W \cdot Z)$$

Substituting in this expression the given values, we have:

$$[0 + (0 \cdot 1 + 1 \cdot 1)] \cdot (1 \cdot 1 + 0 \cdot 1) =$$
$$[0 + (0 + 1)] \cdot (1 + 0) =$$
$$[0 + (1)] \cdot (1) = 1 \cdot 1 = 1$$

Thus, with these conditions there is a complete path, i.e., the expression represents a closed circuit. A summary of these relationships follows:

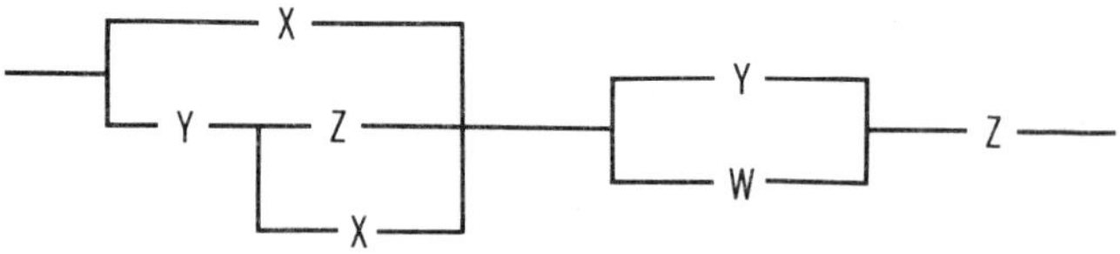

$$X + Y \cdot (X+Z) \cdot (Y+W) \cdot Z = X + (X \cdot Y + Y \cdot Z) \cdot (Y \cdot Z + W \cdot Z)$$

Fig. 3-9. Application of principles to circuits.

$1 = \text{closed}$

$0 = \text{open}$

$1 + 1 = 1$

$1 + 0 = 0 + 1 = 1$

$0 + 0 = 0$

$0 \cdot 0 = 0$

$1 \cdot 0 = 0 \cdot 1 = 0$

$1 \cdot 1 = 1$

$\overline{0} = 1$

$\overline{1} = 0$

$1 \cdot X = X$

$0 \cdot X = 0$

$\cdot = \text{series} = \text{AND}$

$+ = \text{parallel} = \text{OR}$

$X + Y = Y + X$

$X \cdot Y = Y \cdot X$

$X + (Y + Z) = (X + Y) + Z$

$X \cdot (Y \cdot Z) = (X \cdot Y) \cdot Z$

$X \cdot (Y + Z) = X \cdot Y + X \cdot Z$

$X + Y \cdot Z = (X + Y) \cdot (X + Z)$

$X + \overline{X} = 1$

$X \cdot \overline{X} = 0$

$0 + X = X$

$\overline{\overline{X}} = X$

EXAMPLES OF CIRCUIT DESIGN

As an example of a circuit synthesis, in Fig. 3-10A there are two switches, X and Y, arranged so that they control load L in

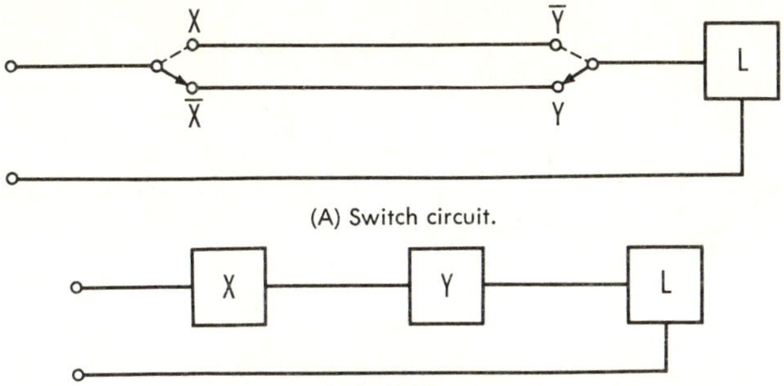

(A) Switch circuit.

(B) Block diagram.

Fig. 3-10. Circuit synthesis.

such a manner as to turn the load *either* on or off. The load is the light; it is controlled by two wall switches. There are only four possibilities:

Case	L	X	Y
a.	0	1	1
b.	1	1	0
c.	1	0	1
d.	0	0	0

Since there is an output (light) for case b and for case c, as well as no output for case a and case d, it is possible to write:

$$L = \overline{X} \cdot Y$$
$$L = X \cdot \overline{Y}$$

The first equation says, in effect, that the load is energized when NOT X (X open) is in the circuit with Y. Since a parallel connection is impossible, the switches must be in series, so $L = \overline{X} \cdot Y$. Thus, if $L = \overline{X} \cdot Y$, or if $L = X \cdot \overline{Y}$, the load is energized. Since the load is energized in case of either expression, the combination of both can only be an OR circuit. Thus we have:

$$L = \overline{X} \cdot Y + X \cdot \overline{Y}.$$

Switches X and Y each have two positions, and there are two possible paths for L to be connected to; $\overline{X} \cdot Y$ is one, and the other is $X \cdot \overline{Y}$. Observe that if $\overline{X} \cdot \overline{Y}$ is used, or if $X \cdot Y$, the load is not connected. Fig. 3-10B is an equivalent block diagram of the switch circuit.

Making the simplification of circuits is easy. Fig. 3-11A shows a circuit for which the expression is:

$$X \cdot Y \cdot (X + Z) \cdot (Y + X \cdot Z + W) \cdot T$$

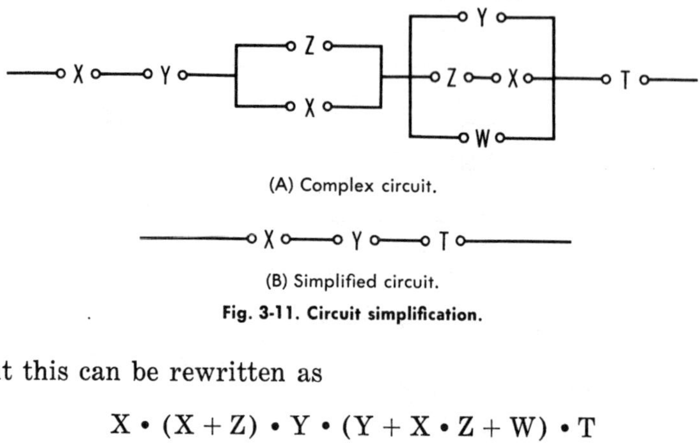

(A) Complex circuit.

(B) Simplified circuit.

Fig. 3-11. Circuit simplification.

But this can be rewritten as

$$X \cdot (X + Z) \cdot Y \cdot (Y + X \cdot Z + W) \cdot T$$

And since $X \cdot (X + Z) = X$, and $Y \cdot (Y + X \cdot Z + W) = Y$, the equivalent expression is $X \cdot Y \cdot T$, for which the circuit is shown in Fig. 3-11B.

Compare the equivalent circuits in Figs. 3-11A and B. It is clear that X, Y, or T individually can break, or open, the circuit. But it is necessary that all three be closed for the circuit to be complete. Neither Z nor W have any effect on the circuit, since they are not in B. This is true for any series contacts. But notice that Z and W are in parallel with other switches. For example, the $Z \cdot X$ portion of the circuit is in parallel with the Y switch. Since, in another part of the circuit, X is in series and is not in parallel with other switches, X must be closed $(X = 1)$ for a complete circuit. And if $X = 0$, the value of Z does not affect the complete path; whether it is closed $(Z = 1)$ or open $(Z = 0)$, there is current flow through $X = 1$ in parallel.

This analysis provides a circuit equivalence that appears unusual at first. Fig. 3-12 illustrates three circuits and their equivalents. The state of X determines the state of the circuit in Fig. 3-12A, since $X + X \cdot Y = X$. The expression for the circuit in Fig. 3-12B is:

(A) Circuit equals X.

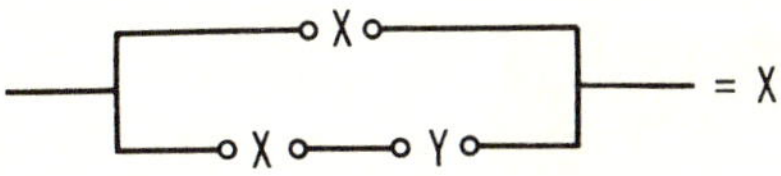

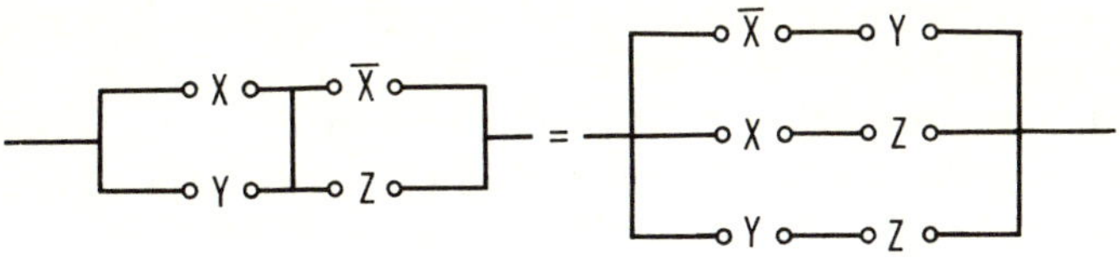

(B) Circuit equals $\overline{X} \cdot Y + X \cdot Z + Z$.

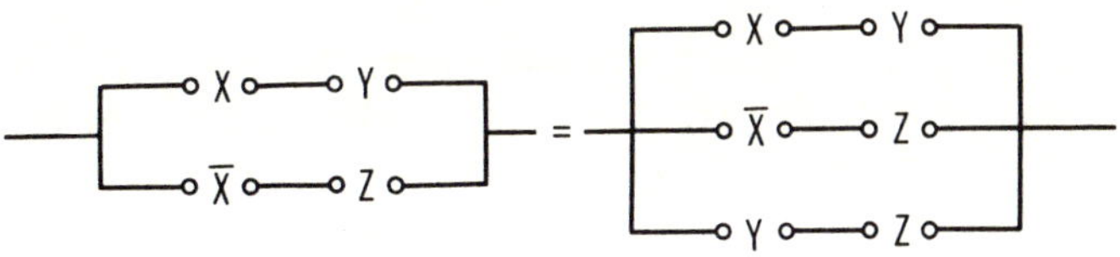

(C) Circuit equals $X \cdot Y + \overline{X} \cdot Z + Z$.

Fig. 3-12. Examples of equivalent logical circuits.

$$(X + Y) \cdot (\overline{X} + Z) = X \cdot \overline{X} + X \cdot Z + \overline{X} \cdot Y + Y \cdot Z$$

But $\overline{X} \cdot X = 0$. Thus the expression reduces to:

$$\overline{X} \cdot Y + X \cdot Z + Y \cdot Z$$

From Fig. 3-12C, we see that:

$$X \cdot Y + \overline{X} \cdot Z = X \cdot Y + \overline{X} \cdot Z + Y \cdot Z$$

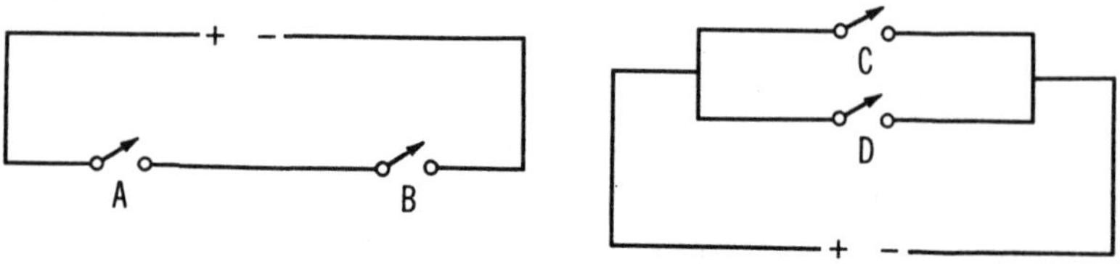

Fig. 3-13. Simple switch circuits.

Multiple-Contact Switching

Relay switching is similar to computer switching and is, in fact, its predecessor. Industrial controls have used switching relays for years, but the development of digital computers led to a deeper analysis of relay switching. The circuits discussed here are all used in industrial control, but they are closely related to those in computer switching. Even now relay contacts are usually drawn to illustrate simple switching circuits.

Fig. 3-13 shows a simple group of circuits, but from these many others may be devised. A and B are in series; both must be closed to complete the circuit, but each can open the circuit. In computer logic this is an AND circuit; current flows only if A and B are both closed. In machine controls, A could be the main power switch and B the operator's foot-switch. When the contacts C and D are in parallel as shown, closing either C or D (this is an OR circuit) permits current. Here both may be closed and the circuit still works; but one alone is not enough to open the complete circuit. Both must be open.

In Fig. 3-14 there are two relays shown. In Fig. 3-15A, 1 and 2 complete the circuit. Fig. 3-14B shows 1 breaking the circuit. Note that 1 can turn the complete circuit either on or off (Fig.

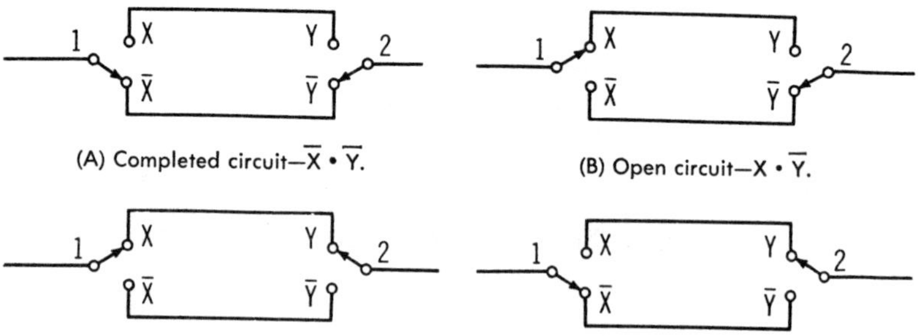

(A) Completed circuit—$\bar{X} \cdot \bar{Y}$.

(B) Open circuit—$X \cdot \bar{Y}$.

(C) Completed circuit—$X \cdot Y$.

(D) Open circuit—$\bar{X} \cdot Y$.

Fig. 3-14. A relay AND circuit.

3-14A or B). Fig. 3-14C shows the next "on" position, and Fig. 3-14D shows the other "off" position. This is also given below where U means up, and D means down.

	X	**Y**	**Circuit**
(A)	D	D	closed
(B)	U	D	open
(C)	U	U	closed
(D)	D	U	open

From this it is clear that both switches *must* be in the same position for a complete circuit. If up is used for X and Y and down for $\overline{X}$ and $\overline{Y}$, the circuit is closed for Fig. 3-14A which is $\overline{X} \cdot \overline{Y}$ and closed for $X \cdot Y$, but open for $\overline{X} \cdot Y$ and $X \cdot \overline{Y}$.

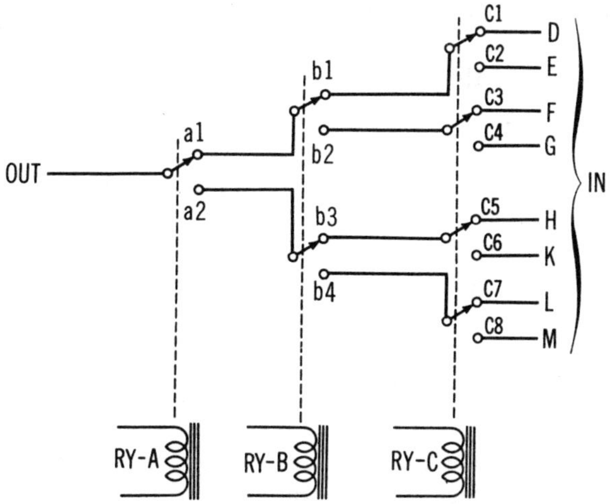

RY-A	RY-B	RY-C	
			D
		✓	E
	✓		F
	✓	✓	G
✓			H
✓		✓	K
✓	✓		L
✓	✓	✓	M

✓= ENERGIZED

Fig. 3-15. A "Christmas tree" relay circuit.

It is only a slight extension to the Christmas-tree relay circuit in Fig. 3-15. Here there are eight inputs; they are controlled so that one (and *only* one) at a time is available at the output. Three relays (A, B, and C) control the seven sets of relay contacts; the output is switched by energizing the proper combination of relays.

If, as illustrated, all relays are open and each contact is in the upper position, the input is at D. RY-C is a relay that moves each C contact down when it is energized. If RY-C is energized, and if RY-A and RY-B are open, the only input is at E. For any given input there is a unique combination of the positions of the relays. For input M, all three relays must be energized. For input K, only relays RY-A and RY-C are energized. In Fig. 3-15 there are three relays for eight inputs. The relationship between the number of relays and the number of inputs is:

No. of Relays	Inputs Possible
1	2
2	4
3	8
4	16
5	32
6	64
7	128

From this table the formula is $2^N = C$, where N is the number of relays and C is the number of inputs. One use of this type of switching occurs when a number of readings are presented on a single remote indicator, and the reading that is available depends on how the control relays are energized. Another, of course, is the representation of alphanumeric information within a digital (binary based) computer.

Adding positions to each relay increases the possibilities for more complex switching. In Fig. 3-16 there are two sets of ganged stepping-relay contacts, C-D and E-F. The power source is represented by G, and the two loads are A and B. Here the problem is to control the loads by both relays so that load B, loads A and B, or neither load is connected to the source. There are nine possibilities, as shown below the circuit. These are listed as cases a, b, c, d, e, f, g, h, and k. A list of the steps of the system's operation follows:

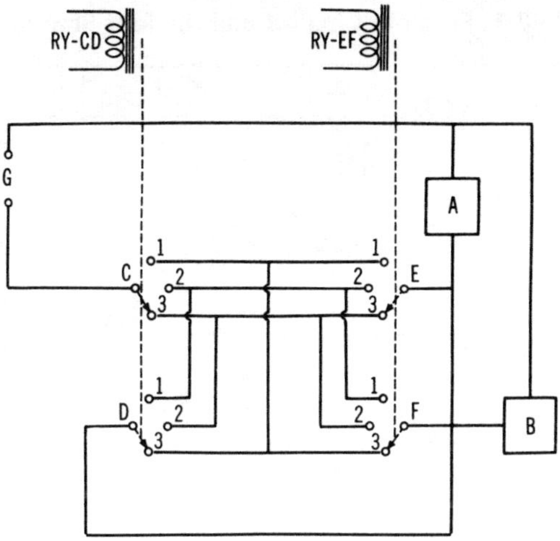

CASES	CD	EF	A	B
c	1	1	ON	ON
b	2	1	OFF	ON
a	3	1	OFF	OFF
f	1	2	OFF	OFF
g	1	3	OFF	ON
k	2	2	ON	ON
h	2	3	OFF	OFF
d	3	2	OFF	ON
e	3	3	ON	ON

NOTE:
+ NO JUNCTION
┬ JUNCTION

Fig. 3-16. Relay switching circuit.

1. Starting with case a, both A and B are not energized; they are open. Using only relay CD, (EF is fixed at 1), first B is closed (b), then both A and B are closed (c). Note that this is by the rotation of CD from 3 to 2 to 1.
2. Starting with case a, both A and B are open. With CD now fixed in position 3, rotation of EF only will go through exactly the same sequence (1, 2, 3), so that the three choices are both A and B, B alone, or neither.
3. Starting with f, both A and B are open. CD is fixed at 1. EF changes to c, which is both A and B closed, or g, which is B closed and A open.
4. 4, 5, and 6 in Table 3-1 show the other possibilities. In 4, CD is fixed; in 5 and 6, EF is fixed.

Table 3-1. Possible Circuit Combinations Shown in Fig. 3-16

		CD	EF	A	B
1	a	↑↓ 3	1	0	0
	b	2	1	0	1
	c	1	1	1	1
2	a	3	↑↓ 1	0	0
	d	3	2	0	1
	e	3	3	1	1
3	c	1	↑↓ 1	1	1
	f	1	2	0	0
	g	1	3	0	1
4	h	2	↑↓ 3	0	0
	k	2	2	1	1
	b	2	1	0	1
5	f	↑↓ 1	2	0	0
	k	2	2	1	1
	d	3	2	0	1
6	g	↑↓ 1	3	0	1
	h	2	3	0	0
	e	3	3	1	1

Note: 0 = open
1 = closed
↑↓ indicates rotation as 1, 2, 3 or 3, 2, 1

Sequence switching with simple relays is indicated in Fig. 3-17. There is a dc power source, a push-button P, and four relays A, B, C, and D. Here it is required that the operating sequence shall be as listed:

Steps	Relays On
1	none
2	A
3	A, B
4	A, B, C
5	A, B, C, D

The operating sequence is as follows:

1. P is open, no relay energized (Fig. 3-17A).
2. P is closed. A is energized, closing a1 and completing a path B, and B is shorted out and does not energize (Fig. 3-17B).

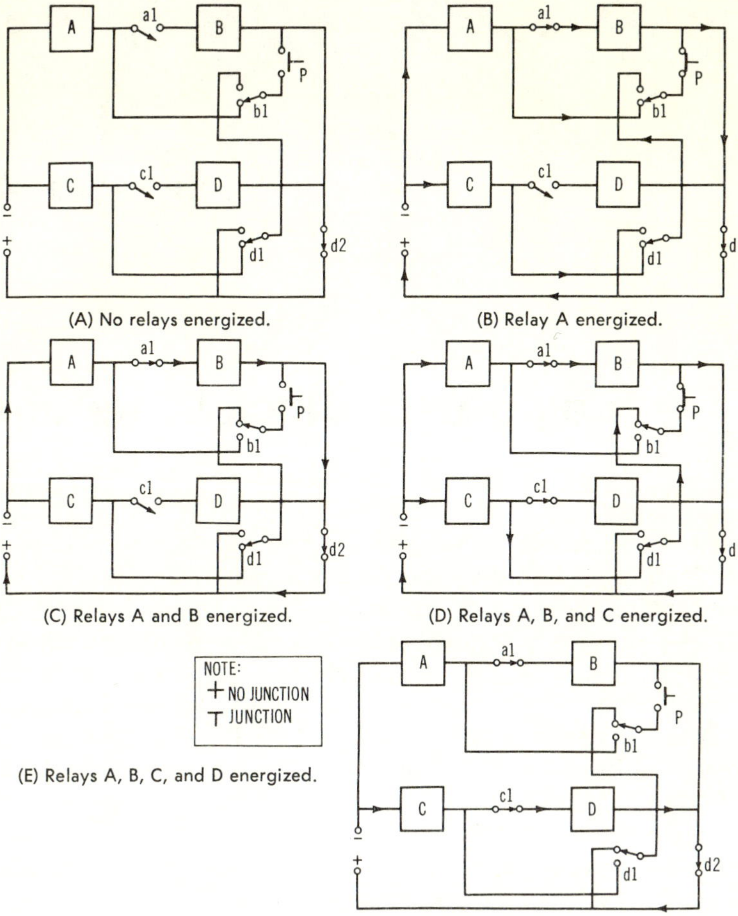

(A) No relays energized.

(B) Relay A energized.

(C) Relays A and B energized.

(D) Relays A, B, and C energized.

(E) Relays A, B, C, and D energized.

Fig. 3-17. Sequence switching with simple relays.

3. P is opened. The short across B is removed, A and B are both energized and they are in series, b1 is in the upper position (Fig. 3-17C).
4. P is closed. C is energized through d1 lower and b1 upper. Both A and B are also energized (Fig. 3-17D). Also c1 is closed, but D is shorted out by the path that energized C.
5. P is opened. The short across D is removed, C and D remain in series and are both energized, d1 is in the upper position, A, B, and C are also energized (Fig. 3-17E). When D is energized, however, it also opens d2, which restores the circuit to the original condition.

CHAPTER 4

Logic With Block Diagrams

The basic logical connectives are AND and OR; they find wide application in digital circuit diagrams and are useful for understanding these circuits in terms of their inputs and outputs. Fig. 4-1 shows a practical application of these connectives to a simple relay circuit. Two relays (M1 and M2) are shown in two arrangements. Each relay has two contacts labeled 1 and 2.

To carry power from source E to load R1, the relays can be connected in series as in Fig. 4-1A. This circuit will pass current only if *both* relays are closed, only if M1 *and* M2 are closed. In Boolean algebra this is the same thing as saying that each condition must be true for the combined statement to be true. Should one condition be false, the circuit will be incomplete and the output will be false. If "true" is represented by a 1 and "false" is represented by a 0, the truth values for all possible conditions are shown in the table accompanying Fig. 4-1A.

In Fig. 4-1B, the relays are connected in parallel. If either relay is closed, current will flow, i.e., if M1 *or* M2 is closed, the circuit will be complete, as represented in its truth table.

Logical symbols reflect the logical meaning of the circuit, regardless of the type of circuit, and regardless of whether transistors, diodes, or gas tubes are used. These are shown in Fig. 4-2; Fig. 4-2A is A AND B, Fig. 4-2B is A OR B, Fig. 4-2C is the inverter producing $\overline{A}$ from A. These may be combined for other logical functions as shown in Fig. 4-3, where the negation of the OR is NOT (A OR B) and the negated AND produces $\overline{A \cdot B}$.

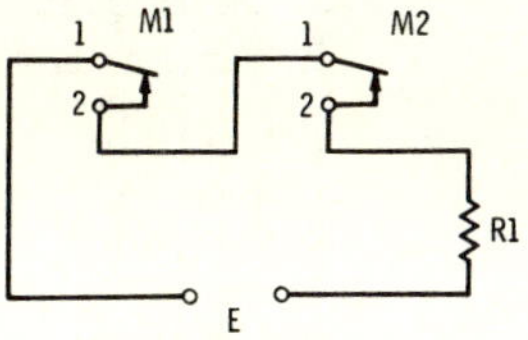

AND Truth Table

M1	M2	CIRCUIT
1	1	1
0	1	0
1	0	0
0	0	0

(A) AND (series) circuit.

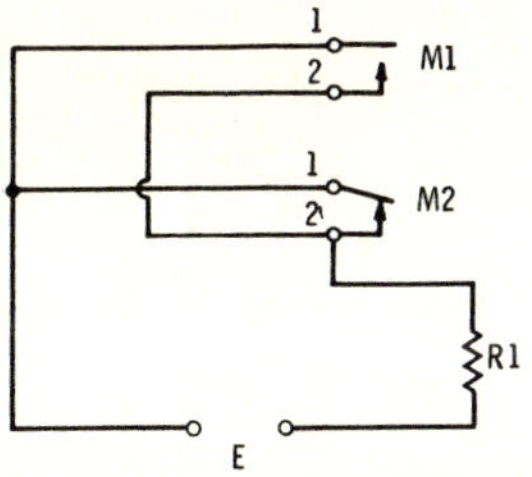

OR Truth Table

M1	M2	CIRCUIT
1	1	1
0	1	1
1	0	1
0	0	0

(B) OR (parallel) circuit.

Fig. 4-1. AND and OR circuits and tables.

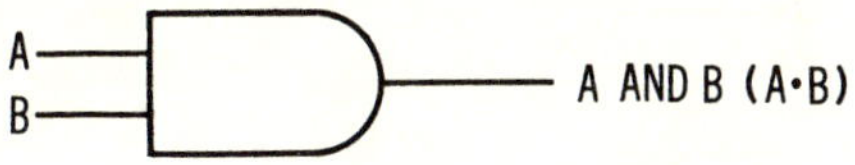

(A) AND.

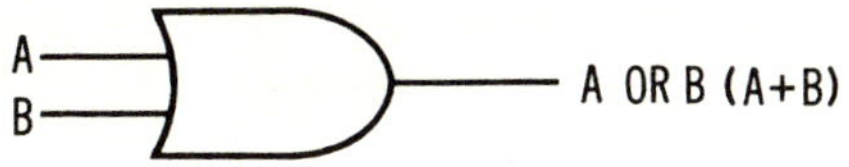

(B) OR.

(C) Inversion.

Fig. 4-2. Logical symbols.

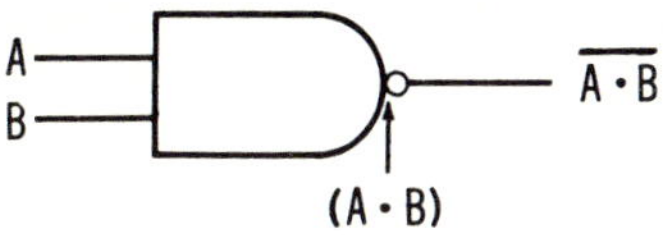

(A) NOT AND.

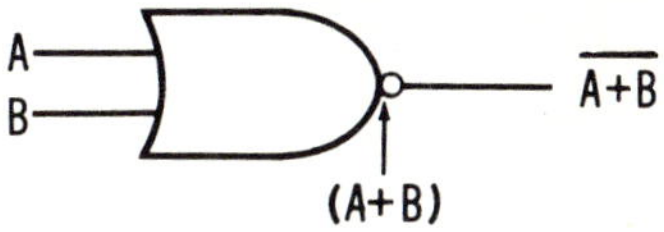

(B) NOT OR.

Fig. 4-3. Negation of logical functions.

Two of the most significant logical functions are the NAND (NOT AND) and NOR (NOT OR).

Consider the three-input NOT AND shown in Fig. 4-4A; the output is low only if all inputs are high. Actually there are two ORs: Fig. 4-4B shows the inclusive OR (negated), where the out-

INPUT			OUTPUT
A	B	C	F
L	L	L	H
L	L	H	H
L	H	L	H
L	H	H	H
H	L	L	H
H	L	H	H
H	H	L	H
H	H	H	L

(A) AND—negated.

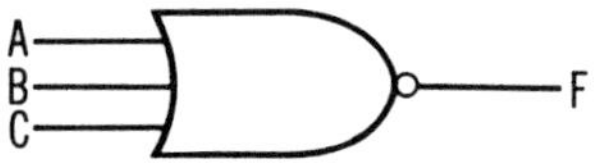

INPUT			OUTPUT
A	B	C	F
L	L	L	H
L	L	H	L
L	H	L	L
L	H	H	L
H	L	L	L
H	L	H	L
H	H	L	L
H	H	H	L

(B) Inclusive OR—negated.

INPUT		OUTPUT
A	B	F
L	L	H
L	H	L
H	L	L
H	H	H

(C) Exclusive OR—negated.

Fig. 4-4. Logical negation.

put is low if any one of the inputs is high; Fig. 4-4C shows the exclusive OR, where the output is low if only one input is high.

Fig. 4-5, shows eight combinations relating the AND and OR functions. In Fig. 4-5A, for example, $A \bullet B = X$ is the same as $\overline{A} + \overline{B} = \overline{X}$, which is an expression of De Morgan's theorem.

LOGICAL BLOCKS

Other circuit blocks are available for building logical systems; some of these are described in this section.

The flip-flop shown in Fig. 4-6A is a device that stores a single bit of information. It has three possible inputs, set (S), reset (C), and trigger (T), and two possible outputs, 1 and 0. Reset is sometimes called clear.

The two outputs are normally of opposite polarity. A 1 is stored in the flip-flop when the 1 output level is active, and the

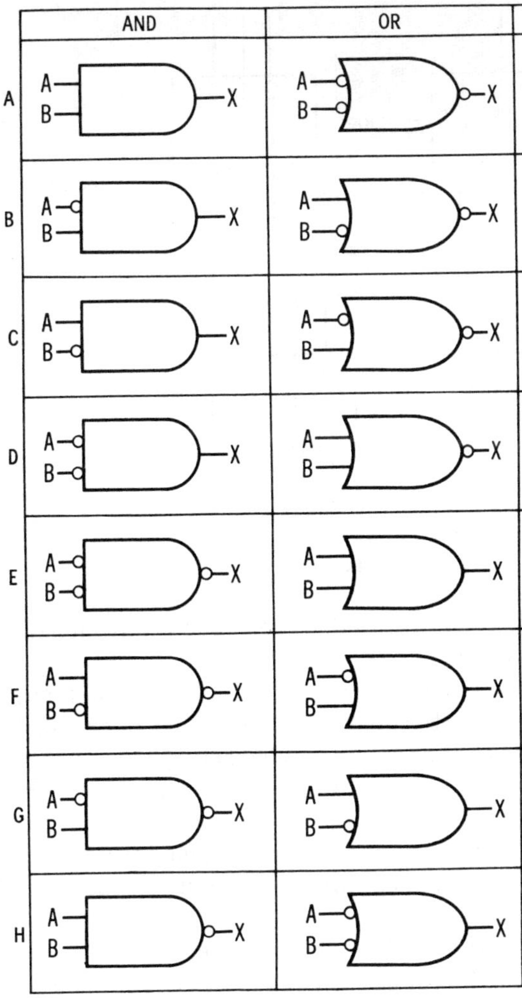

Row	A	B	X
A	H	H	H
	H	L	L
	L	H	L
	L	L	L
B	H	H	L
	H	L	L
	L	H	H
	L	L	L
C	H	H	L
	H	L	H
	L	H	L
	L	L	L
D	H	H	L
	H	L	L
	L	H	L
	L	L	H
E	H	H	H
	H	L	H
	L	H	H
	L	L	L
F	H	H	H
	H	L	L
	L	H	H
	L	L	H
G	H	H	H
	H	L	H
	L	H	L
	L	L	H
H	H	H	L
	H	L	H
	L	H	H
	L	L	H

Fig. 4-5. Combinations of AND and OR functions.

0 output level is inactive. A 0 is stored in the flip-flop when this condition is reversed. The flip-flop assumes the 1 state when an active signal appears at the S input, regardless of the original state. It assumes the 0 state when an active signal appears at the "C" input, regardless of the original state. It reverses its state when an active signal appears at the T input. There are several possible variations to normal flip-flop operations, depending on the response of the device when active inputs are

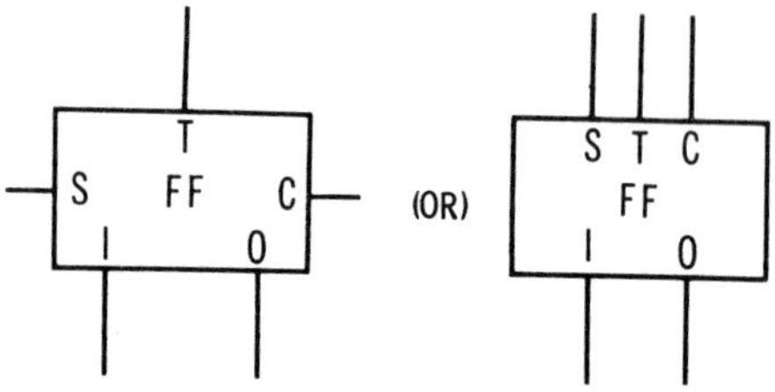

(A) Flip-flops.

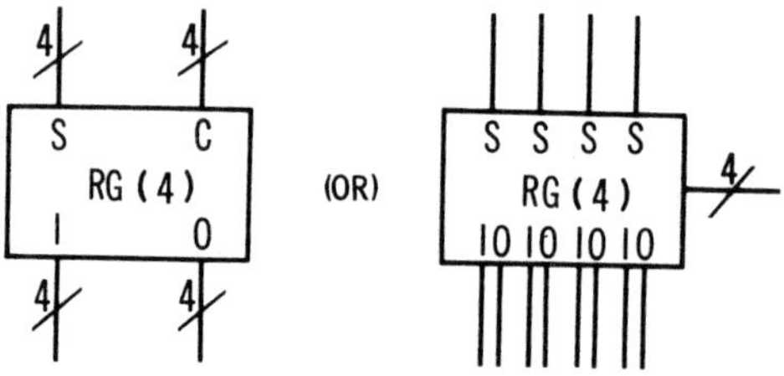

(B) Binary register.

Fig. 4-6. Logical blocks.

simultaneously applied. The S input is near the 1 output; the C input is near the 0 output.

The binary register symbol (Fig. 4-6B) represents a group of flip-flops used in parallel to constitute a single register (such as would be used to store four bits of a character). It is necessary to indicate the number of bits or individual flip-flops in the register. Examples show four S inputs grouped on one multiple-input line, and four pairs of 1 and 0 grouped output lines. In some applications, individual input and output lines are shown, as in the right hand figure.

The *shift-register* symbol (Fig. 4-7) represents a binary register with provision for displacing or shifting the content

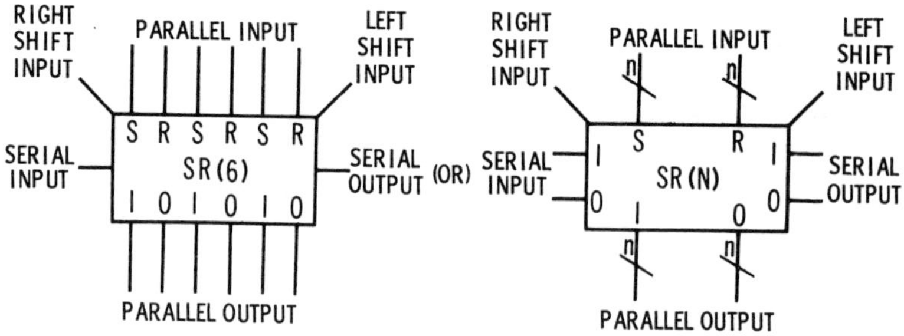

Fig. 4-7. Shift register symbol.

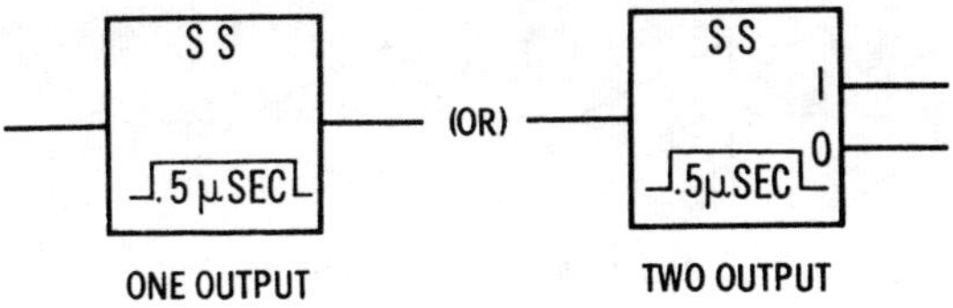

Fig. 4-8. Single-shot symbol.

of the register one stage at a time; it is shifted to the right or left by means of the shift input. The words "right shift input" are usually placed at a left corner of the symbol to indicate a shift from left to right. If the shift is from right to left, the words "left shift input" are placed at a right corner of the symbol.

Fig. 4-8 shows the symbol representing single-shot (SS) functions. Output signal shape, amplitude, duration, and polarity are determined by the circuit characteristics of the SS, (not by the input signal) and may be shown inside or outside the symbol. The quiescent state of the SS is either zero or one. When actuated, it changes to the opposite state and remains in that state for a specified time dependent on the design of the device.

Fig. 4-9 represents the Schmitt Trigger (ST) function. This device is actuated when the input signal exceeds a threshold voltage. Output signal amplitude and polarity are determined by the circuit characteristics of the ST (not by the input signal). Stylized waveforms may be shown (inside or outside the symbol), indicating amplitude, polarity, threshold voltage, and duration. The quiescent state of the ST is either 0 or 1. When actuated, it changes to the opposite state and remains in that state as long as the input exceeds the threshold value.

Fig. 4-10 shows a linear or nonlinear, current or voltage amplifier. This amplifier may have one or more stages and can produce either gain or inversion. Level changers and inverters, pulse amplifiers, emitter followers, cathode followers, relay and

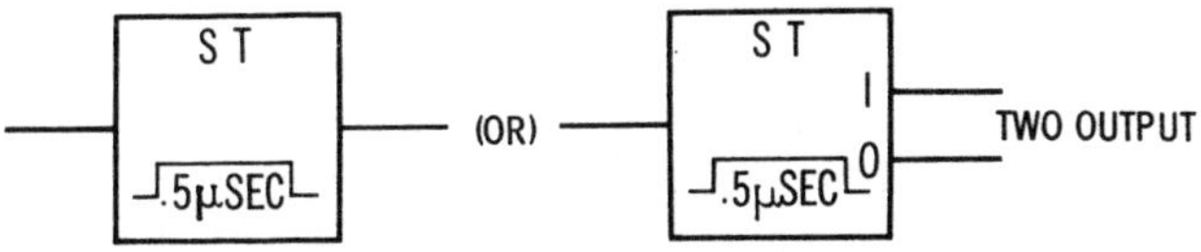

Fig. 4-9. Schmitt trigger symbol.

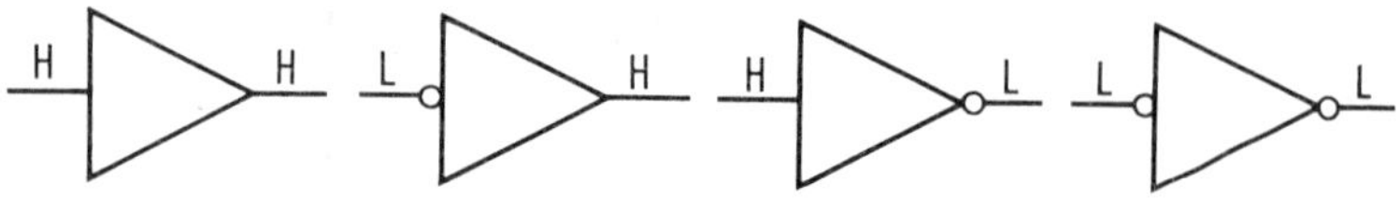

Fig. 4-10. Amplifier symbols.

lamp drivers, and shift register drivers are devices represented by this symbol.

A *time delay* is shown in Fig. 4-11. The duration of the delay is included with the symbol. If the delay device is tapped, the delay time with respect to the input is included adjacent to the tap output. Twin vertical lines indicate the input side.

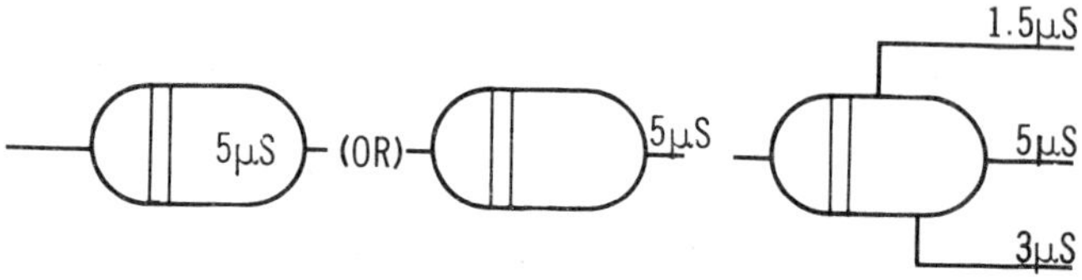

Fig. 4-11. Time delay symbol.

LOGICAL LEVELS OF ACTIVITY

The AND and OR functions are duals: a single arrangement of circuits may perform both the AND function and the OR function. This functional quality is employed in numerous single-device and multidevice systems. We may consider the AND function as an element whose output is active when all its inputs are active. Any nonactive AND input produces a nonactive output. The OR function is considered an element whose output is active when any one or more inputs are active. When all OR inputs are nonactive, the circuit produces a nonactive output.

To identify the activity of a device selected to implement the logic, the state condition of active inputs and the resultant active outputs are identified by active state signal indicators (small circles) at the inputs or outputs of logic functions (AND/OR). These graphic representations, as well as the English notations, illustrate the relationship of specific functions. A small circle at the inputs indicates that the relatively low (L) input signal activates the function. Conversely, the absence of a small circle indicates that the relatively high (H) input signal activates the function. A small circle at the symbol output side indicates that the output of the activated function is relatively low. Absence of a small circle at the symbol output indicates that the output of the activated function is relatively high.

Activating inputs, or an activated output of a function, may be:

1. Logical 1 in either the high state (H) or the low state (L).
2. A logical 0 either high or low.
3. A mixture of both 1 or 0 either high or low.

Consider a device whose active output (F) is a function of two signals (A, B). The output and both input levels are capable of assuming only the arbitrarily chosen values, +2 volts (H) and −3 volts (L). The circuit behaves according to Device Activity State Table A of Table 4-1. Substitution of the abbreviation H for the +2 volt levels and L for the −3 volt levels results in Activity Combinations Table C (Table 4-1).

When the +2 volt level is considered the activating level and is assigned the logic value 1, and the −3 volt level is considered

Table 4-1. Activity States for Output = F (A, B)

(Electrical Truth)
Device Activity States
Table A

Input		Output
A	B	F
+2 volts	+2 volts	+2 volts
+2 volts	−3 volts	−3 volts
−3 volts	+2 volts	−3 volts
−3 volts	−3 volts	−3 volts

AND Function Activity States
Table B

Input		Output
A	B	F
1	1	1
1	0	0
0	1	0
0	0	0

Activity Combinations
Table C

Input		Output
A	B	F
H	H	H
H	L	L
L	H	L
L	L	L

OR Function Activity States
Table D

Input		Output
A	B	F
0	0	0
0	1	1
1	0	1
1	1	1

the inactive level and is given the logic value 0, then substitution of these logic state values for the Table A active voltage levels results in AND Function Activity States (Internal) Table B (Table 4-1). The device is now said to perform the AND function. When the −3 volt level is the activating level and is assigned the logic value 1 and the +2 volt level is considered the inactive level and has the logic value 0, then substitution of these logic state values for Table A active voltage levels results in OR Function Activity States (Internal) Table D (Table 4-1). The device is now said to perform the OR function.

Consider a different device whose active output (F) is a function of two signals (B, C). The output and both input levels are capable of assuming only the arbitrarily chosen values +2 volts (H) and −3 volts (L). The circuit behaves according to Device Activity States Table A of Table 4-2.

Table 4-2. Activity States for Output = F (B, C)

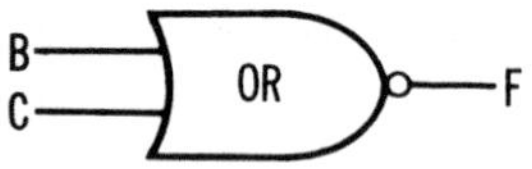

(Electrical Truth)
Device Activity States
Table A

Input B	Input C	Output F
−3 volts	−3 volts	+2 volts
−3 volts	+2 volts	−3 volts
+2 volts	−3 volts	−3 volts
+2 volts	+2 volts	−3 volts

Activity Combinations
Table B

Input B	Input C	Output F
L	L	H
L	H	L
H	L	L
H	H	L

AND Function Activity States
Table C

Input B	Input C	Output F
1	1	1
1	0	0
0	1	0
0	0	0

OR Function Activity States
Table D

Input B	Input C	Output F
0	0	0
0	1	1
1	0	1
1	1	1

Substitution of mnemonic abbreviation H for +2 volt levels and L for –3 volt levels in Device Activity States Table A results in Activity Combinations Table B (Table 4-2).

Inputs (B, C) –3 volt levels are the activating input levels and are assigned the logic value 1; the +2 volt output level (F) is considered the activated output and is also assigned the logic value 1; inactive input levels +2 volts and the inactive –3 volt output level are assigned the logic value 0. Substitution of these logic state assignments for Table A circuit voltage levels results in AND Function Activity States Table C. The device is now said to perform the AND logic function defined by Table C and the AND inverting operation defined by Table A and combinations Table B. The device is symbolized by combining the AND function symbol with input level indicators (less positive than F).

Substitution of abbreviations H for +2 volt levels and L for –3 volt levels results in Activity Combinations Table B where H is high and L is low. Inputs (B, C) +2 volt levels are considered the activating input levels and are assigned the logic value 1, and the –3 volt output level (F) is considered the activated output and is also assigned the logic value 1; inactive input level –3 volt and the inactive +2 volt output level are assigned to the logic value 0. Substitution of these logic state assignments for Table A circuit voltage levels results in OR Function Activity States Table D (Table 4-2). The device is now said to perform the OR logic function as in Table D and the OR inverting input operation as defined by Table A and combinations Table B. The device is symbolized by combining the OR function symbol with an output level indicator (less positive than B, C).

Electrical state English notations are added to signal line inputs and outputs for identification when that line is either logical 1 or 0 in a logic network of operations. For example, if line P (H) is placed at the input to logic elements, notation (H) indicates that line P signal is high, that is, a logical 1, when it exists. If upon inspection, line input P (H) is low, then it is in the logical 0 state. This nonactive logical 0 low state output can activate a device input in the logic network. Table 4-3 illustrates this concept.

As in Fig. 4-12 a given line signal P (H) can be active or inactive depending on the point under discussion. When P (H) is

Table 4-3. Electrical State Notations

Activity State	Device States	
	+2 volts	−3 volts
P (H)	1	0
B (H)	1	0
T (L)	0	1

high, as noted, it is in the logical 1 state and will produce output X but will inhibit output Y. Conversely, when P does not exist as a high, it is in the logical 0 state for the OR function and produces AND output Y if line T is low.

A given signal must be considered and, when necessary, notated in terms of three independently variable parameters for every point in the logic network. These parameters are:

1. Logical State; presence (1) or absence (0).
2. Electrical State; high or low.
3. Activity State; Signal Line condition, noted by graphic representation (presence or absence of small circles) or English notations (line named high or low).

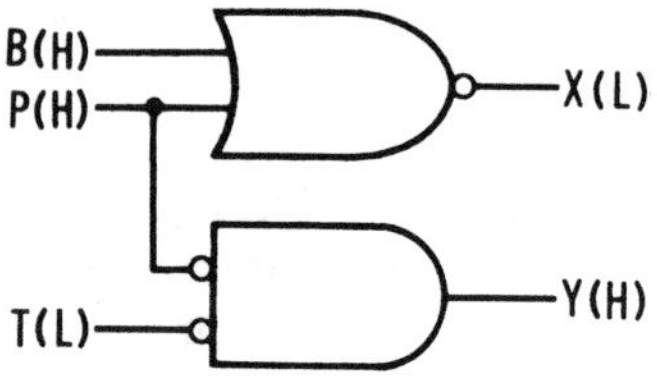

X = (B + P)(L)

B (H) or P (H) or both B (H) and P (H)

Y = (P • T) (H)

P (L) and T (L) = P and T (H)

Fig. 4-12. Logical network.

LOGICAL OPERATIONS

Involved operations are more easily understood with the proper combination of these blocks. The following block diagram arrangements may be rearranged by the use of the logical relationships. These rearrangements are not obvious from the original. In each case, the more simple and direct equation will serve the same function as the original from which it is derived.

To show that $A + \overline{A} \cdot B$ is the same as $A + B$, as in Fig. 4-13:

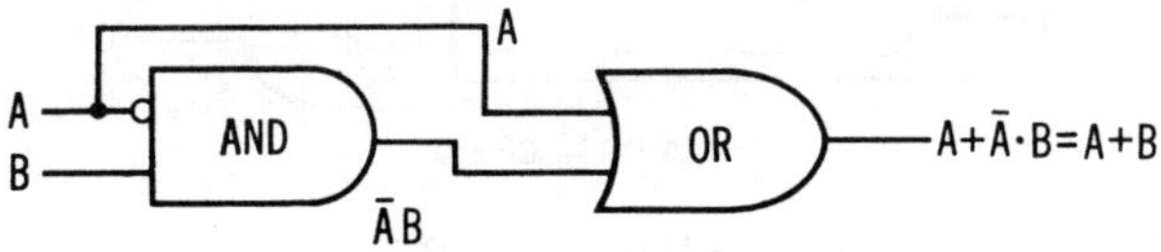

Fig. 4-13. AND and OR combination.

$$\begin{aligned}A + \overline{A} \bullet B &= A \bullet (B + \overline{B}) + \overline{A} \bullet B, \text{ since } B + \overline{B} = 1\\ &= A \bullet B + A \bullet \overline{B} + \overline{A} \bullet B\\ &= A \bullet B + A \bullet \overline{B} + A \bullet B + \overline{A} \bullet B\\ &= A \bullet (B + \overline{B}) + B \bullet (A + \overline{A})\\ &= A + B\end{aligned}$$

Another example is illustrated in Fig. 4-14. We want to show:

$$A \bullet B + A \bullet C + B \bullet \overline{C} = A \bullet C + B \bullet \overline{C}$$

Now,

$$\begin{aligned}A \bullet B + A \bullet C + B \bullet \overline{C} &= A \bullet B\,(C + \overline{C}) + A \bullet C + B \bullet \overline{C}\\ &= A \bullet B \bullet C + A \bullet B \bullet \overline{C} + A \bullet C + B \bullet \overline{C}\\ &= A \bullet C \bullet (B + 1) + B \bullet \overline{C} \bullet (A + 1)\\ &= A \bullet C + B \bullet \overline{C}\end{aligned}$$

Combinations of AND gates and OR gates are shown in Fig. 4-15. Fig. 4-15A is $A \bullet B + C = D$; Fig. 4-15B is $A \bullet B \bullet C + D \bullet E + F \bullet G = H$.

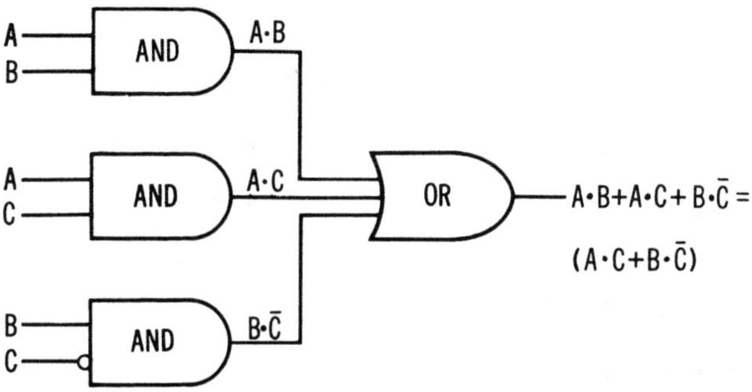

Fig. 4-14. Logical operation circuit.

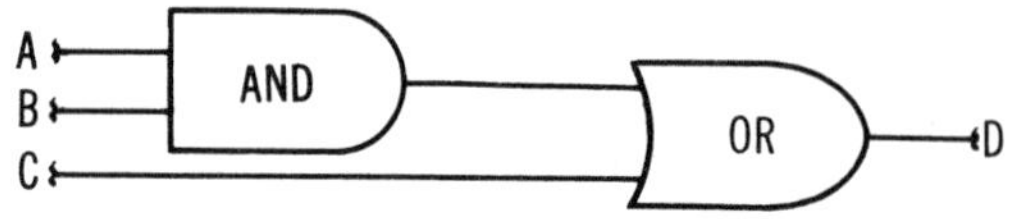

(A) $A \cdot B + C = D$.

The Boolean expression $A \cdot B \cdot C + D \cdot E + F \cdot G = H$ *may be expressed symbolically as follows:*

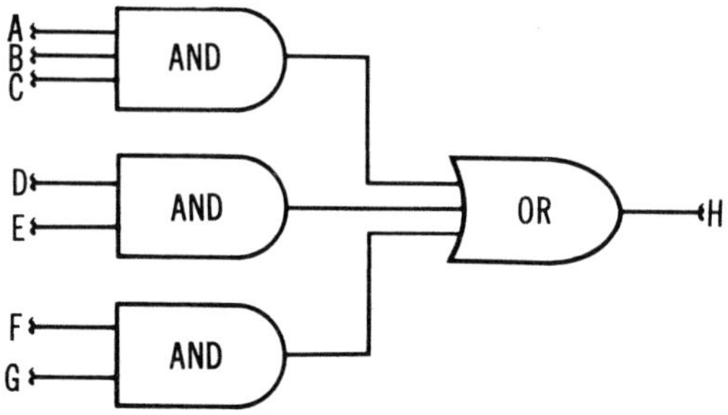

(B) $A \cdot B \cdot C + D \cdot E + F \cdot G = H$.

Fig. 4-15. Logical circuit arrangements.

Fig. 4-16 shows $A \cdot (B + C) + D \cdot E + F \cdot G \cdot H \cdot I = K$. This requires three two-legged AND gates and one four-legged AND gate and one four-legged OR gate.

As another example, the relationship $(A + B) + (A \cdot C + B)$ can be reduced by Boolean algebra. This expression can be simplified to:

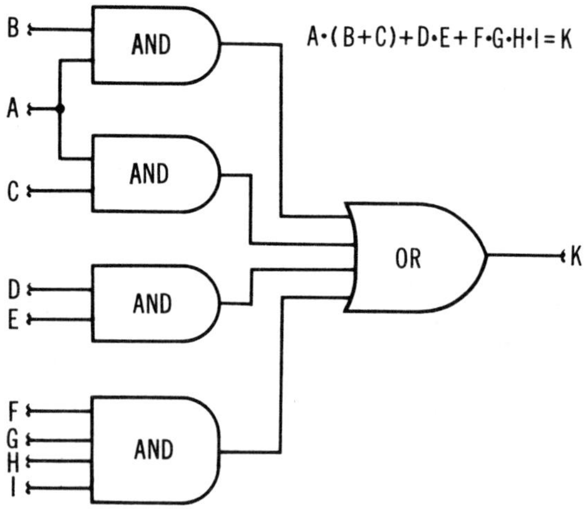

Fig. 4-16. Circuit implementing complex expression.

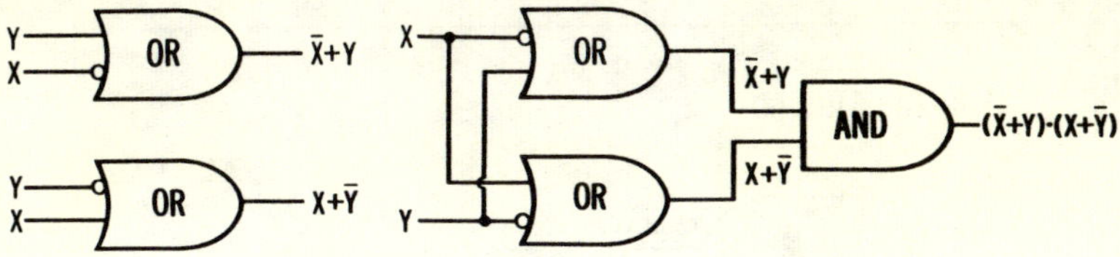

Fig. 4-17. How diagrams are developed from expressions.

$$
\begin{aligned}
(A + B) + (A \bullet C + B) &= (A + A \bullet C) + (B + B) \\
&= (A + A \bullet C) + B \\
&= A + B
\end{aligned}
$$

Therefore, a simple OR gate can be used.

Diagrams can be developed directly from expressions as (NOT X OR Y) AND (X OR NOT Y). To draw the block diagram for $(\overline{X} + Y) \bullet (X + \overline{Y})$, each expression is first set up. The two expressions are then combined as shown in Fig. 4-17. We may also change relationships from AND to OR. The change from AND to OR (and from OR to AND), from $(\overline{A \bullet B \bullet C})$ to $\overline{A} + \overline{B} + \overline{C}$, as in Fig. 4-18, may be made by the use of De Morgan's theorem.

NOR/NAND LOGIC

For implementing binary algebra, two logic assignments, NOR and NAND, are possible. In one system, NOR, a logical 1 is represented by the presence of a voltage and a logical 0 is represented by no voltage. The NAND system results if the assignments are reversed.

	Binary 0	Binary 1
NOR	Low voltage	High voltage
NAND	High voltage	Low voltage

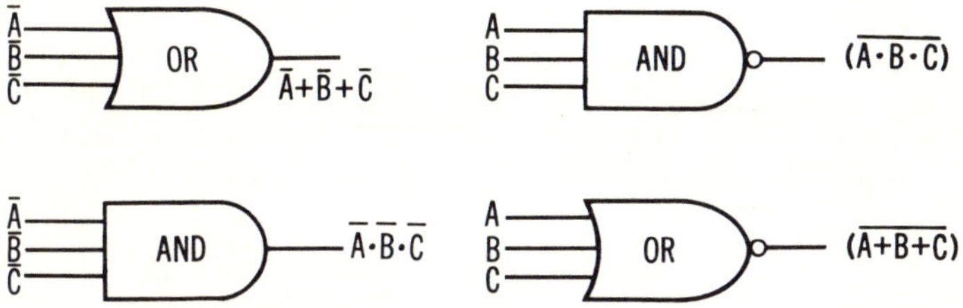

Fig. 4-18. Implementation of De Morgan's Theorems.

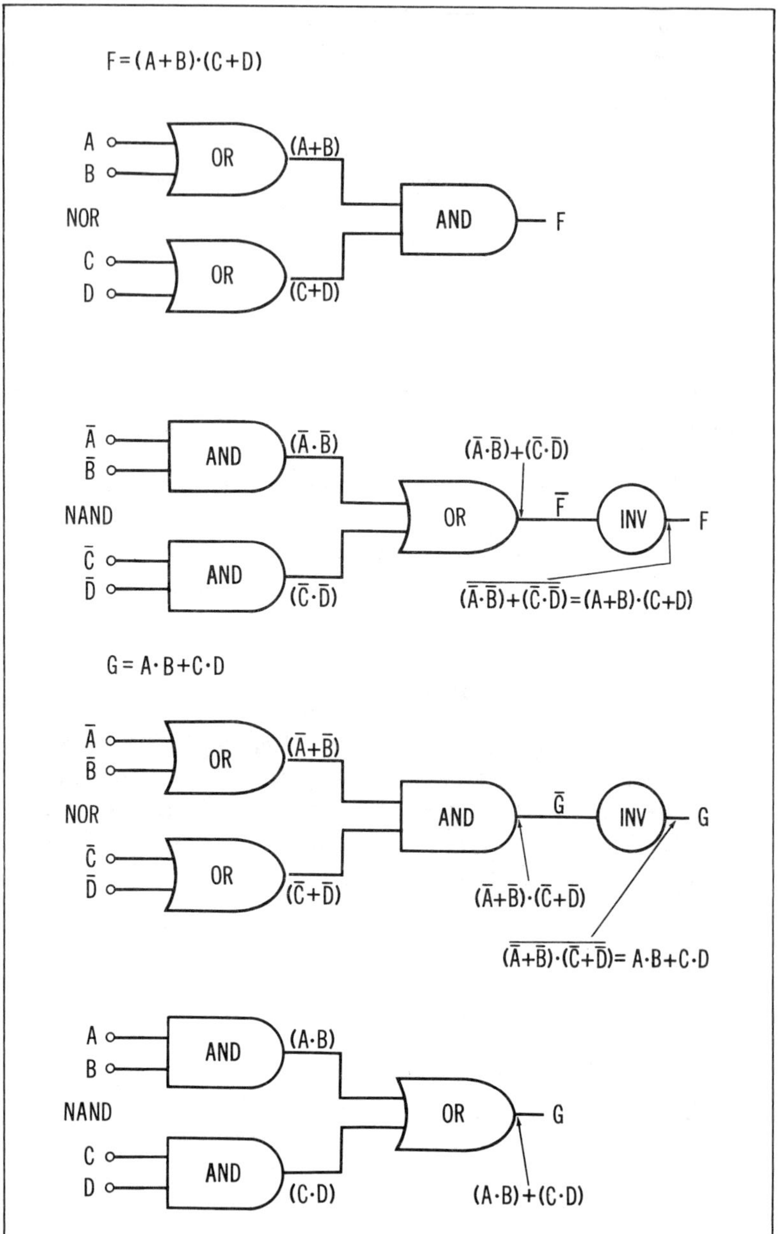

Fig. 4-19. NOR/NAND logical circuits.

A gate can be used as either a NOR or NAND gate, depending upon the logic system selected. The transfer functions of the gates are:

NOR	A, B, . . . , X	$\overline{A} \bullet \overline{B} \cdots \overline{X}$ or $\overline{A + B + \cdots + X}$
NAND	A, B, . . . , X	$\overline{A} + \overline{B} + \cdots + \overline{X}$ or $\overline{A \bullet B \cdots X}$

$F = (A + B) \bullet (C + D) = A \bullet C + B \bullet C + A \bullet D + B \bullet D$

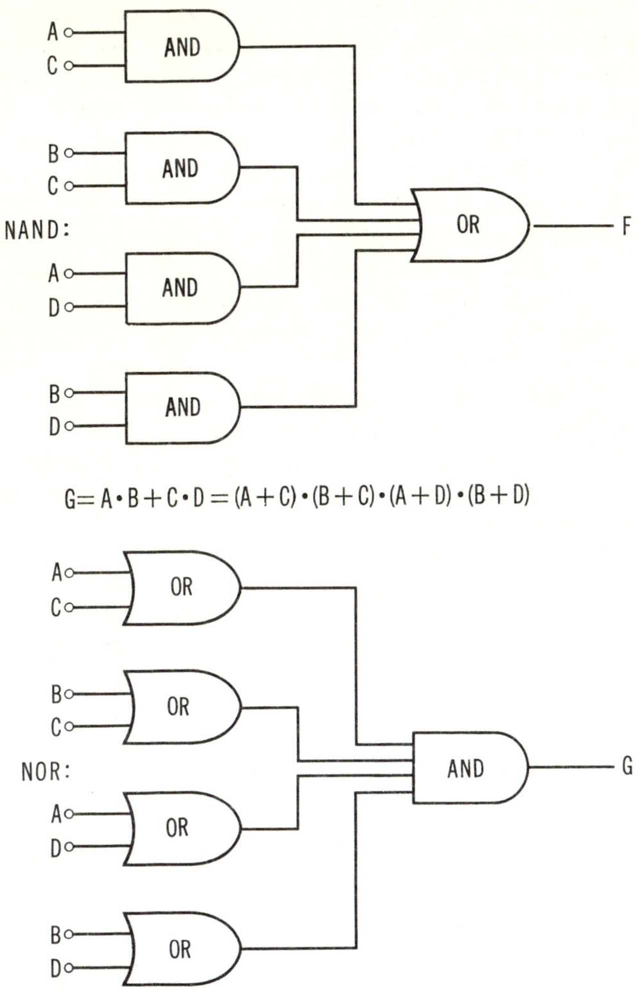

Fig. 4-20. NOR/NAND duality.

Fig 4-19 shows the duals $F = (A + B) \cdot (C + D)$ and $G = A \cdot B + C \cdot D$.

Consider using the two equations shown below:

$$F = (A + B) \cdot (C + D) = A \cdot C + B \cdot C + A \cdot D + B \cdot D$$

$$G = A \cdot B + C \cdot D = (A + C) \cdot (B + C) \cdot (A + D) \cdot (B + D)$$

As in Fig. 4-20, it can be seen that the two logic systems are duals with the NOR logic favoring maxterm type equations and the NAND logic favoring minterm type equations. Employing a particular equation with its favored type of logic ordinarily results in the best system with respect to economy and speed. It is sometimes possible to reduce the number of gates connected in series, hence increasing speed at the expense of using extra networks. In the previous example, the two equations may be transformed to achieve this purpose. Any equation may be readily reduced to its simplest minterm form, but it is often difficult to expand an equation that is in minterm form into a maxterm type equation that is reduced to the optimum form for implementation. For this reason, NAND logic is usually preferred.

Combinations or NOR and NAND logic are shown in Fig. 4-21. Fig. 4-22 is a summary of logic symbols.

All basic logic symbols drawn without small inversion circles at the input or output are considered to operate as positive

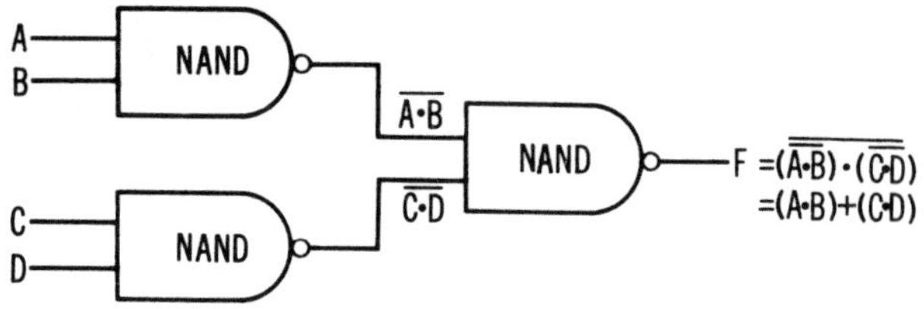

(A) NAND logic.

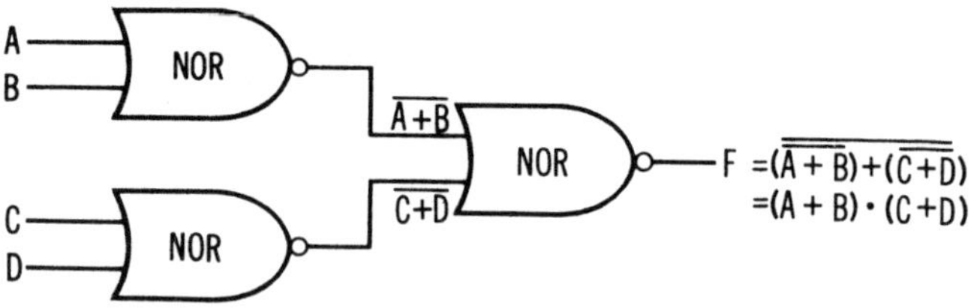

(B) NOR logic.

Fig. 4-21. Combinations of NAND and NOR logic.

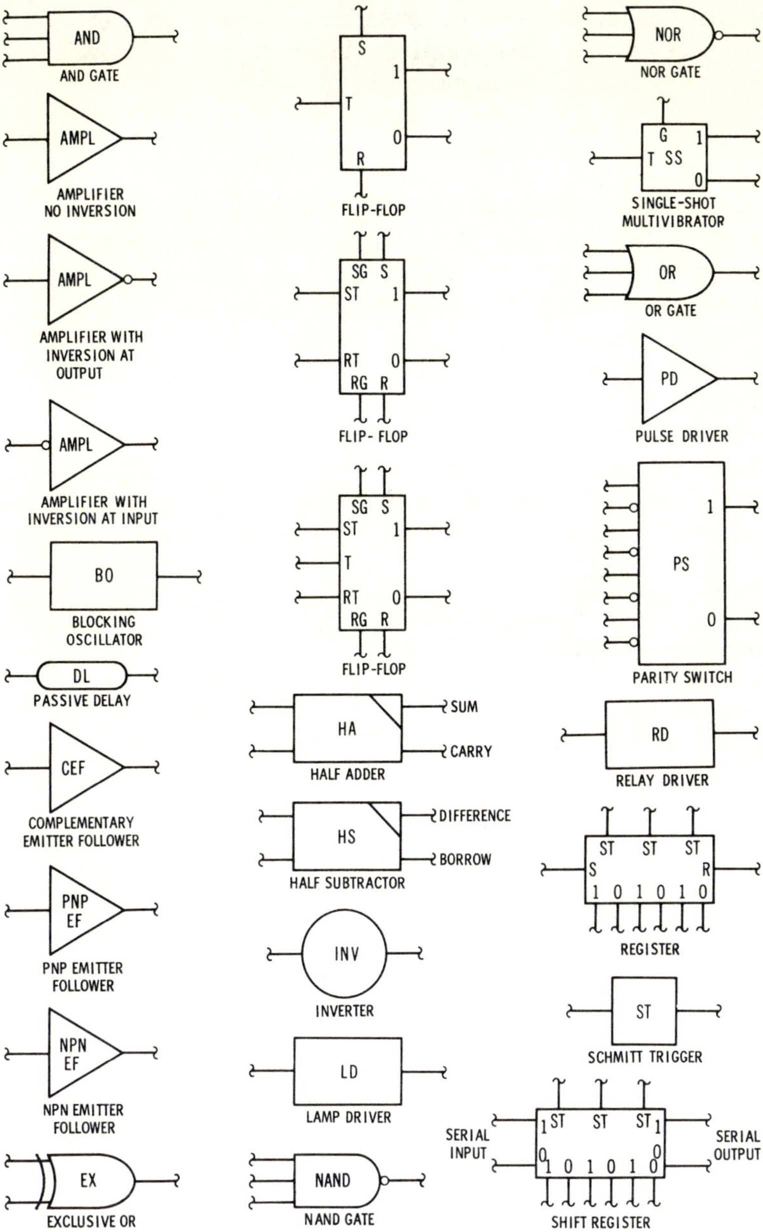

Fig. 4-22. Diagrammatic symbols used in logic circuits.

logic elements. Positive logic is designated as the most positive relative dc level that is equivalent to high state, true, or binary 1. Therefore, the most negative relative dc level is equivalent to low state, false, or binary 0.

COMPUTER CIRCUITRY

Computing circuitry generally requires various forms of arithmetic operations. These operations are usually performed with fundamental devices called either half-adders or half-subtracters, grouped in appropriate combinations.

The addition of two binary numbers generates a sum bit, either 0 or 1 depending on the addition result, and a carry bit identical to the carry in normal or decimal addition. A full-adder, composed of two half-adders, is required to accommodate a full sum and carry operation for each binary bit. Fig. 4-23 illustrates a circuit and the associated truth table for the generation of the half-addition of two binary bits. All logic

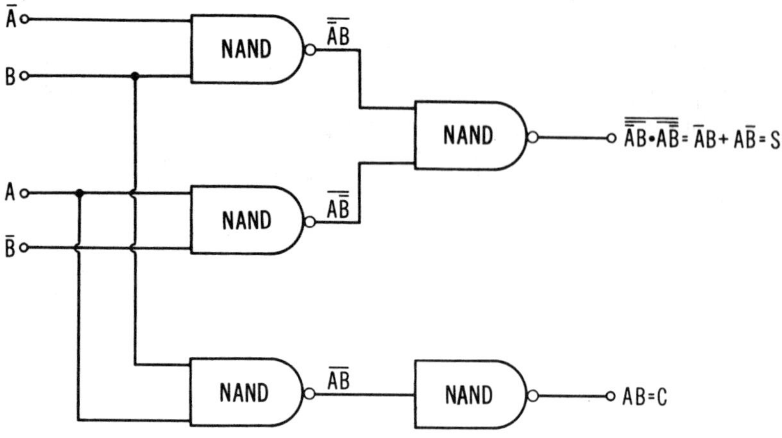

A	B	SUM S	CARRY C
0	0	0	0
0	1	1	0
1	0	1	0
1	1	0	1

THUS: $S = \bar{A}B + A\bar{B}$

$C = AB$

Fig. 4-23. Half-adder circuit.

functions generated are shown on the diagram with the algebraic reduction for the completed equations of the sum and carry bits.

Fig. 4-24 shows the circuitry and truth table for the half-subtraction of two binary bits. The equations for the difference and borrow are similar to those for the sum and carry in a half-adder. Specifically, the sum and difference equations are identical. Only the carry and borrow equations differ. Thus, the major portions of the circuitry for the half-adder and half-subtracter are identical; the only difference is in the generation of the carry or borrow bits.

The full-adder is shown in Fig. 4-25. When using the carry information to correct a serial binary sum, or when a third carry input from a half-adder of the next lowest significant digit is to be added to the present sum, a full-adder must be employed. An example of the logic required for the full adder is shown. The full-adder logic may be simplified by utilizing the exclusive OR logic element as in Fig. 4-26.

The full-subtracter is shown in Fig. 4-27. A full-subtracter may be formed by combining two half-subtracters, an OR gate, and a single-unit delay.

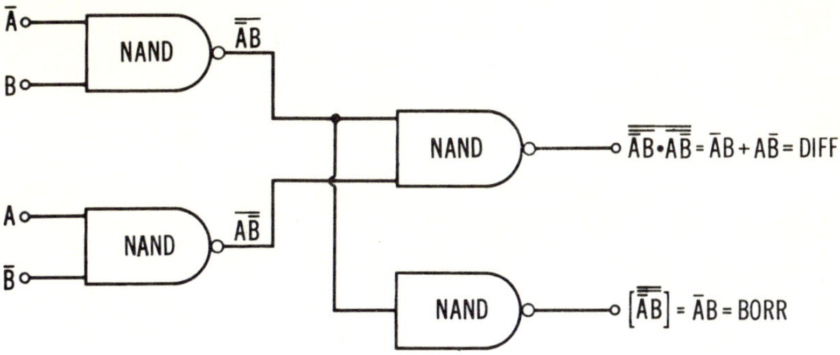

A	B	DIFF	BORR
0	0	0	0
0	1	1	1
1	0	1	0
1	1	0	0

$DIFF = \bar{A}B + A\bar{B}$

$BORR = \bar{A}B$

Fig. 4-24. Half-subtracter circuit.

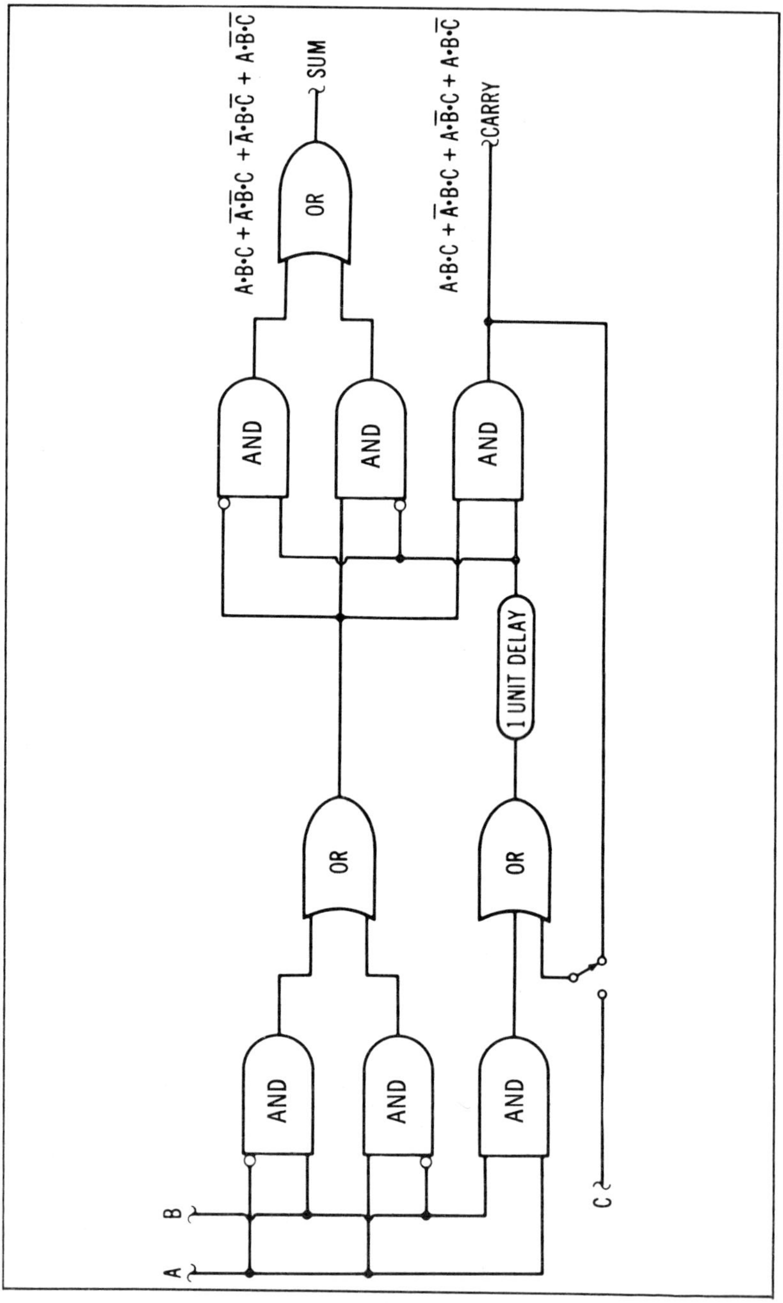

Fig. 4-25. Full-adder circuit.

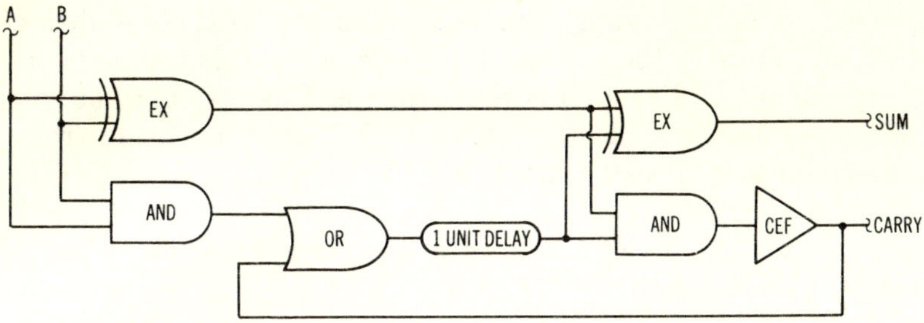

Fig. 4-26. Full-adder logic using exclusive OR.

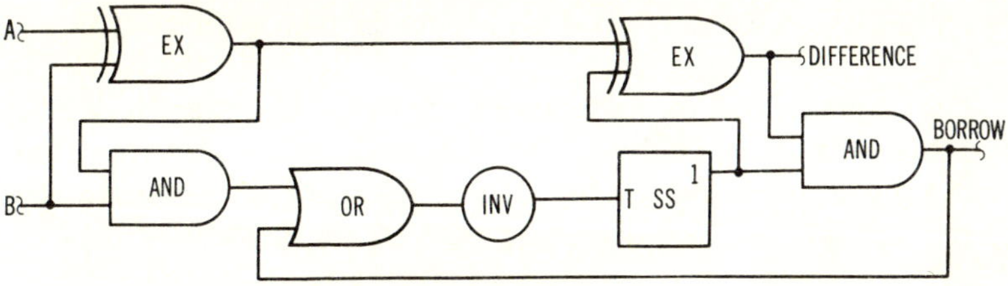

Fig. 4-27. Full subtracter.

As before, the addition of binary numbers is direct and simple when only two digits are to be added. The OR circuit is used, since for a two-input OR, only when both inputs are 0 does a 0 output result. If either input is a 1, the output is a 1. If both inputs are 1's, the output is a logical 1, which represents in this case, a 10 or a carry 1.

Table 4-4. Adding Two-Digit Numbers

	A	B	C	D	E
	Addend	Augend	Carry-in	Sum	Carry-out
1	1	0	0	1	0
2	0	1	0	1	0
3	0	0	1	1	0
4	1	1	0	0	1
5	1	0	1	0	1
6	0	1	1	0	1
7	1	1	1	1	1
8	0	0	0	0	0

Thus, to add two-digit numbers, there are eight possibilities (Table 4-4) : A is the addend, B is the augend, C is the carry-in from the last addition, D is the sum, and E is the carry-out to the next addition. There are three different cases, aside from the trivial case of all 0's. These are one 1, two 1's, and three 1's.

For a single 1, there are three cases: $(A + \overline{B} + \overline{C})$, $(\overline{A} + B + \overline{C})$, and $(\overline{A} + \overline{B} + C)$. There is always a sum and never a carry. For any two 1's, $(A + B + \overline{C})$, $(A + \overline{B} + C)$, and $(\overline{A} + B + C)$, there is always a carry but never a sum. When every digit is a 1, $(A + B + C)$, there is a carry and a sum. When every digit is a 0, $(\overline{A} + \overline{B} + \overline{C})$, there is neither a sum nor a carry.

The implementation may be seen in Fig. 4-28, with the AND's and the OR's as indicated. The INV block is an inverter that changes a 1 to a 0 and a 0 to a 1. A description of the operation of the logical circuits shown in Fig. 4-28 follows.

All 1's OR's 2, 3, 4 each have two 1's, whch means three 1's into AND 2; the 1 into INV means 0 into OR 1. But, 1's through AND 1 and OR 5 means a sum of 1. OR's 2, 3, 4, as with sum, all have 1's, which means AND 2 will have a carry of 1.

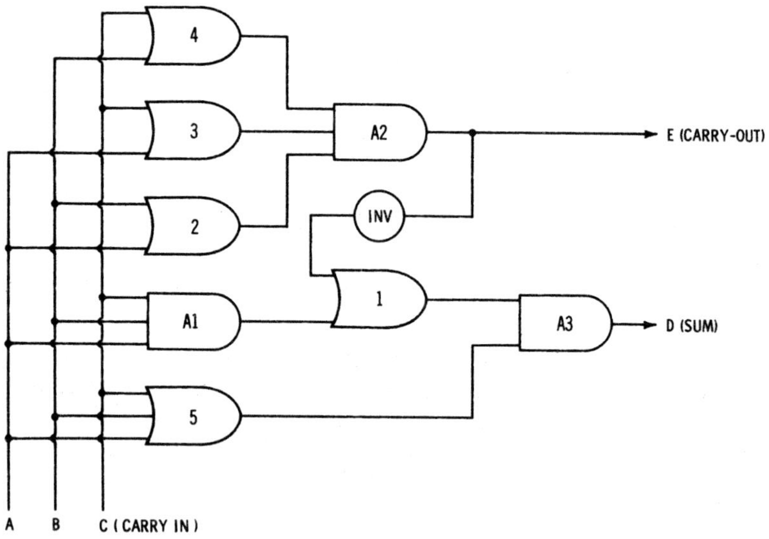

Fig. 4-28. Logic circuit for addition.

Two 1's Any two 1's through OR's 2, 3, 4, will mean a 1 out from AND 2. Through INV, a 0 will be sent to OR 1. But through AND 1 there will be a 0, since all three inputs to AND 1 are not 1; thus there will be no sum since AND C has but a single 1 through OR 5. Since AND 2 has three 1's there will be a carry of 1.

One 1 AND 2 does not have three 1's; thus through INV there is a 1 to OR 1, which passes a 1 to AND 3. OR 5 also has a 1 out because of a single 1 in. AND 3 thus has a sum out of 1. AND 2 has not three 1's for only two OR's (2, 3, 4) can have a 1. Thus there is no carry of 1 but 0.

No 1's Neither AND 1 nor OR 5 has a single 1; hence there is no sum of 1. No 1's into AND 2; hence there can be no CARRY of 1 out, but a CARRY of 0.

CHAPTER 5

The Algebra of Sets

The logician makes a sharp distinction between the two values of a variable in Boolean algebra. He postulates that the range of the variable is a collection of elements, each of which may be tested to meet only one of two conditions; either the element belongs to a predetermined set, or it does not. The mathematical structure of the system of collections is called the *algebra of sets*. To the logician this structure is identical with Boolean algebra.

ELEMENTS AND SETS

We define a *set* to be any collection into a whole of separate and distinct objects. Thus a set is a collection of elements. The elements of a set will be denoted a, b, c, . . . z; the sets will be denoted A, B, C, . . . Z. If the object a belongs to the set A, we write $a \in A$, read "a is a member of A." If A and B are sets, and every element of A is also an element of B, we write $A \subset B$, read "A is a *subset* of B." If $A \subset B$, and there is at least one element of B that does not belong to A, then we say A is a *proper* subset of B. Equality of sets is defined in the obvious way; we say $A = B$ if $A \subset B$ and $B \subset A$. The *universal* set is the set of all elements under discussion and is denoted by 1. The *empty* set is the set having no members and is denoted by 0. The *complement* of a set A is the set of all elements not in A and is denoted A'.

Table 5-1. Symbols of Logic

	OR	AND	NOT	X, Y
Boolean Algebra	X + Y	X · Y	$\overline{X}$	Switches
Propositional Logic	X ∨ Y	X ∧ Y	—X	Propositions
Algebra of Classes	X ∪ Y	X ∩ Y	X′	Sets

Given two sets X and Y, there are various ways of combining them. One way is a union of sets. The *union* of X and Y is the set containing every element of both X and Y. Hence the meaning of union is that of the non-exclusive OR, denoting X OR Y or both. The symbol for the union of two sets X and Y is X ∪ Y. It has the logical meaning of OR (Table 5-1).

The *intersection* of two sets X and Y is the set containing

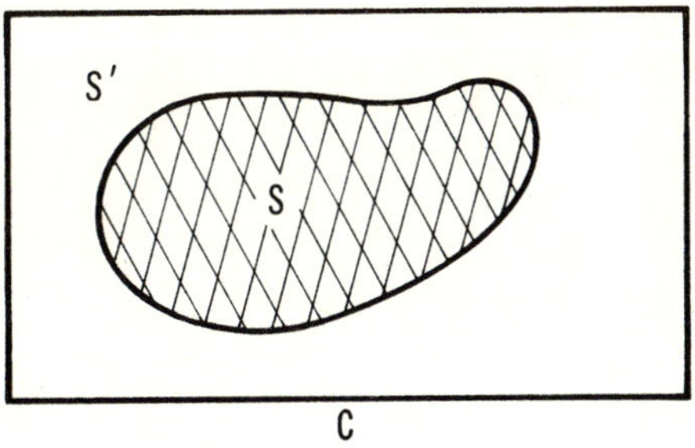

(A) Nonintersecting sets.

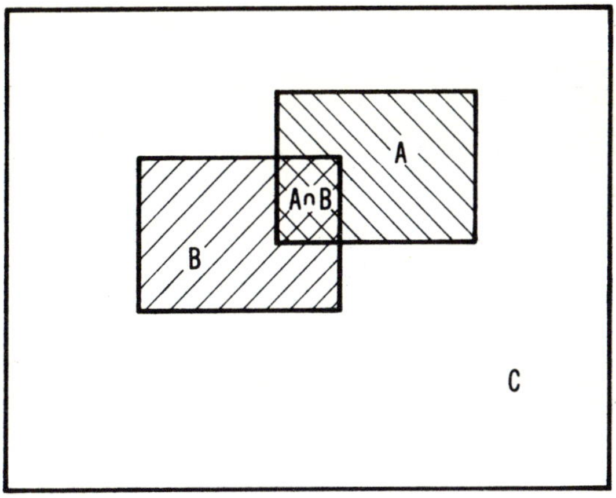

(B) Intersecting sets.

Fig. 5-1. Examples of sets.

every element common to X and Y. In contrast to union, intersection means AND and represents the logical sum of X and Y. The symbol denoting the intersection of two sets X and Y is $X \cap Y$ (Table 5-1).

Consider the sets in Fig. 5-1A. The universal set (1) is the rectangle C. One can show that, for the set S, $S \cup S' = C$, i.e., all elements in S or S′ constitute the universe (C) under discussion. Also observe that $S \cap S' = 0$, i.e., the elements common to S and S′ compose the empty set (that is, there are none).

In Fig. 5-1B there are two rectangles, A and B. It is clear that $A \subset C$, and $B \subset C$, i.e., every element in A is in C, and every element in B is in C. Now consider the set $A \cap B$; all the elements of $A \cap B$ are in the rectangle designated $A \cap B$. But

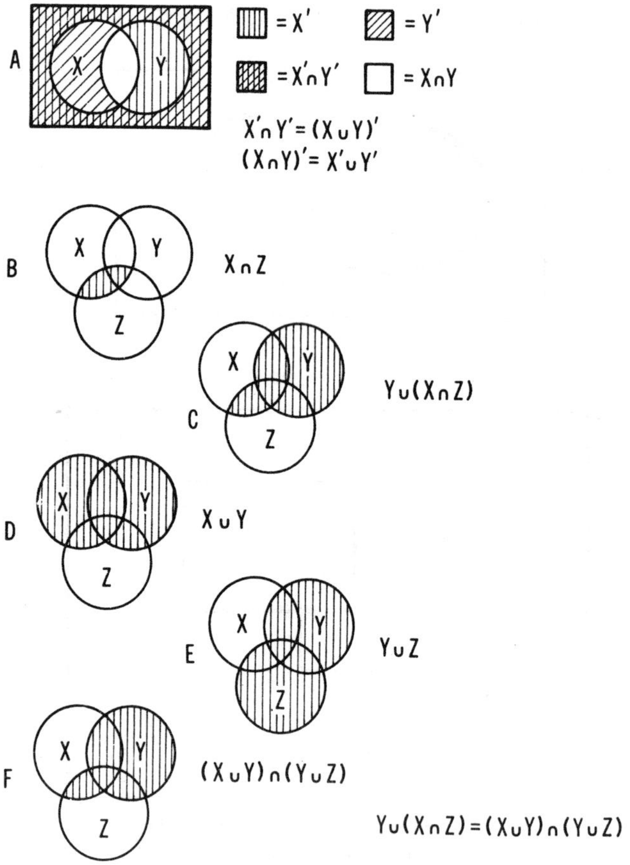

Fig. 5-2. Venn diagrams.

for the union of A and B, the new set $A \cup B$ is the set of all elements in A and in B, and this new set includes the common area of A and B.

Diagrams that illustrate logical relations often are Venn diagrams (Fig. 5-2). In Fig. 5-2A we consider two sets X and Y. The area of vertical shading represents the complement X' of X. Note that the diagram illustrates $X \cap X' = O$, and $X \cup X' = 1$. The area with diagonal shading represents Y'. Note that $X' \cap Y'$ is the area with both vertical and diagonal lines. This area is also $(X \cup Y)'$. Hence, we have $(X \cup Y)' = X' \cap Y'$. Observe that the area common to X and Y is unshaded; this represents $X \cap Y$. It also represents $(X' \cup Y')'$.

If this is so, then:

$$X \cap Y = (X' \cup Y')'$$
$$(X \cap Y)' = X' \cup Y'$$

Consider Fig. 5-2B. The shaded area in this Venn diagram represents $X \cap Z$. In Fig. 5-2C we have included the set Y, and the shaded area is $Y \cup (X \cap Z)$. Fig. 5-2D illustrates the set $X \cup Y$, and Fig. 5-2E illustrates $Y \cup Z$. If we intersect the sets $X \cup Y$ and $Y \cup Z$, we obtain a set:

$$(X \cup Y) \cap (Y \cup Z)$$

that is shown as the shaded area in Fig. 5-2F. Note that the shaded area in this diagram is the same as the area shaded in Fig. 5-2C. Hence we observe that:

$$Y \cup (X \cap Z) = (X \cup Y) \cap (Y \cup Z)$$

and that the Venn diagram is a very useful representation of the union and intersection of sets.

Very important to the concept of number is a relation known as similarity. We say that two sets, A and B, are similar if a 1 to 1 correspondence can be established between the members of A and the members of B. This relation is not the same as equality, since two similar sets may be entirely different collections. For example, if we write [r,s,t,w] to denote the set A consisting of elements r,s,t, and w, and if [c,d,e,f] denotes a set B consisting of elements c,d,e, and f, then we will observe that A and B are similar (Table 5-2). It is important to observe that two similar sets have the same number of elements.

Early in this chapter the concept of the empty set (O) and

Table 5-2. Similarity of Sets

A =	[r, s, t, w]
	↕ ↕ ↕ ↕
B =	[c, d, e, f]

the universal set (1) was discussed. There is also a further characterization of sets, and this defines the difference between the finite and the infinite set. If there is no end to the number of individual members in a set, this set is known as an *infinite* set. For example, the set of all even numbers is infinite. In contrast to this, if the totality of the members of a set can be counted, it is known as a *finite* set. Clearly, the number of members in an infinite set is always greater than the number of members in a finite set. It is possible to show that there are as many numbers as there are even numbers. For example, let A be the set of all whole numbers such as 1, 2, 3, 4, 5, 6, . . . and B be the set of all even numbers, such as 2, 4, 6, 8, 10, 12, ... We obtain the 1 to 1 correspondence simply by pairing each number with its double, e.g., $1 \leftrightarrow 2$, $2 \leftrightarrow 4$, $3 \leftrightarrow 6$, . . . , etc. Thus there are as many even numbers as there are whole numbers.

OPERATIONS ON SETS

In the algebra of sets, there are certain rules that define the manipulation of sets. There are seven basic identities formed by using only the operations of union, intersection, and negation. Those identities are listed for reference:

(1) (a) $0 \cap X = 0$
(b) $1 \cap X = X$
(c) $0' = 1$
(d) $1' = 0$
(e) $0 \cup X = X$
(f) $1 \cup X = 1$
(2) (a) $X \cap Y = Y \cap X$
(b) $X \cup Y = Y \cup X$
(3) (a) $X \cap (Y \cup Z) = (X \cap Y) \cup (X \cap Z)$
(b) $X \cup (Y \cap Z) = (X \cup Y) \cap (X \cup Z)$
(4) (a) $X \cap (Y \cap Z) = (X \cap Y) \cap Z$
(b) $X \cup (Y \cup Z) = (X \cup Y) \cup Z$
(5) (a) $X \cap X = X$

(b) $X \cup X = X$
(c) $X \cap (X \cup Y) = X$
(d) $X \cup (X \cap Y) = X$
(6) (a) $X \cap X' = 0$
(b) $X \cup X' = 1$
(c) $(X')' = X$
(7) (a) $(X \cap Y)' = X' \cup Y'$
(b) $(X \cup Y)' = X' \cap Y'$

The preceding relations could be established using the diagram shown in Fig. 5-3. In this diagram, we have:

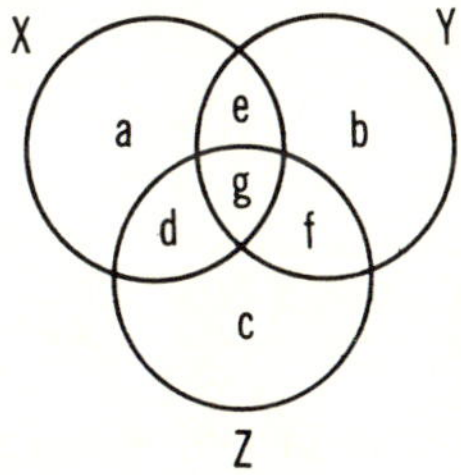

Fig. 5-3. Diagram used to establish the rules for manipulation of sets.

$$X = [a, e, d, g]$$
$$Y = [b, e, g, f]$$
$$Z = [c, d, f, g]$$

For example to find $X \cup Y$:

$$X \cup Y = [a, e, d, g] \cup [b, e, g, f]$$
$$= [a, e, d, g, b, f]$$

And to find $X \cap Y$:

$$X \cap Y = [a, e, d, g] \cap [b, e, g, f]$$
$$X \cap Y = [e, g]$$

We can also use this technique to verify the seven relations previously given.

We can also simplify expressions. For example:

$$X \cap (X' \cup Y) \cup Y \cup [Y \cap (Y \cup Z)]$$
$$= (X \cap X') \cup X \cap Y \cup Y \cup (Y \cap Y) \cup (Y \cap Z)$$

But by (6)a, $X \cap X' = 0$
by (5)b, $X \cup X = X$
by (5)a, $X \cap X = X$

Thus, $(X \cap X') \cup X \cap Y \cup Y \cup (Y \cap Y) \cup (Y \cap Z)$ becomes $(X \cap Y) \cup Y \cup (Y \cap Z)$.

Also, by (5)d, $X \cup (X \cap Y) = X$. Therefore, the expression becomes simply Y.

Applied to sets, De Morgan's law provides a dual relationship between intersection and union. This duality exists in such a way that all unions may be replaced by intersections; all intersections may be replaced by unions; all 1's may be replaced by 0's; all 0's may be replaced by 1's, and the result will still be an identity. Some simple examples are shown in Table 5-3.

Table 5-3. Dual Relation Between Union ($\cup$) and Intersection ($\cap$)

$$\left.\begin{array}{l} 0 \cap X = 0 \\ 1 \cup X = 1 \end{array}\right\}$$

$$\left.\begin{array}{l} 1 \cap X = X \\ 0 \cup X = X \end{array}\right\}$$

APPLICATIONS OF SETS

The algebra of sets has several applications. Consider a line A drawn on a sheet of paper. This line is defined as the intersection of two planes, and the line extends as far as we wish in both directions. This line X may be considered to be composed of an infinite number of points as in Fig. 5-4. If on this line X

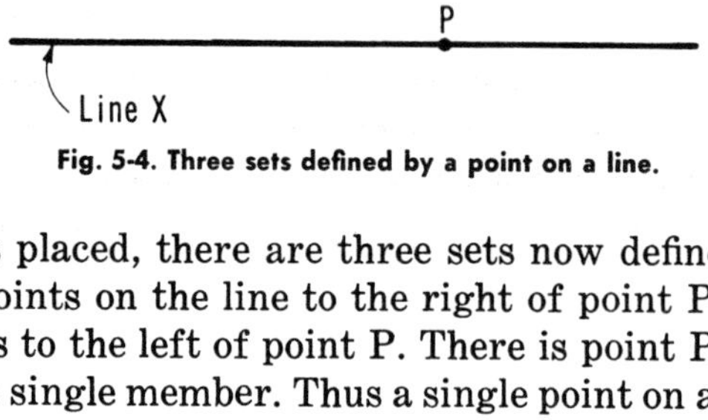

Fig. 5-4. Three sets defined by a point on a line.

a point P is placed, there are three sets now defined. There is the set of points on the line to the right of point P. There is a set of points to the left of point P. There is point P, which is a set having a single member. Thus a single point on a line defines three sets.

A surprising correspondence between two lines is shown in Fig. 5-5. For any point d on line DE, there is a corresponding point on base BC of triangle ABC. To show this, draw a straight line from the vertex A through point d. This line intersects BC in one and only one point b. And conversely, a line drawn from point b to the vertex A will intersect the line DE in one and only one point d. Hence there is a 1 to 1 correspondence between the set of points composing line DE and the set

Fig. 5-5. Two lines having the same number of points.

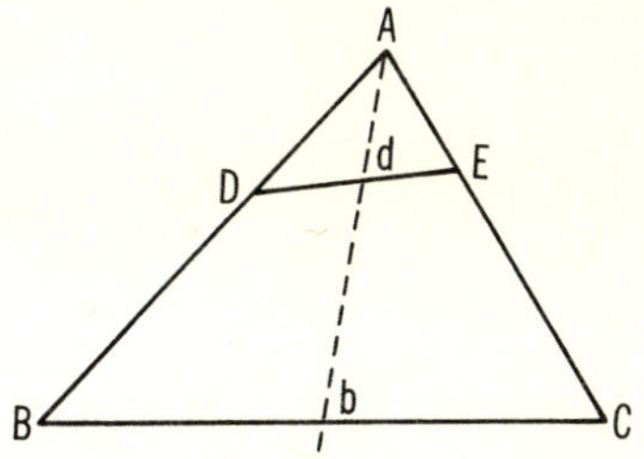

composing line BC. We must conclude, then, that DE has the same number of points as BC.

The expression [x|x < 7] is read as "the set of all numbers x such that x is less than 7." Because this set is infinite, it is not possible to enumerate all of the members; it is, however, possible to graph such a set. Fig. 5-6A shows the graph of this expression or the solution set of x < 7. A rounded arrowhead means the set does *not* include 7, but all numbers up to 7.

Suppose A = [x|x> 4], i.e., A is the set of all numbers x such that x is greater than 4. Also suppose B = [x|x < −2], i.e., B is the set of all numbers x such that x is less than −2. Then if C = A ∪ B, C = [x|x > 4 or x|x < −2], i.e., C is equal to the union of A and B, as in Fig. 5-6B.

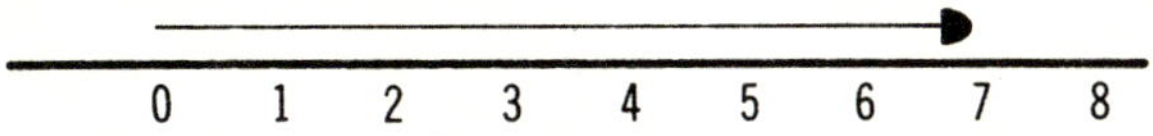

(A) Graph of x < 7.

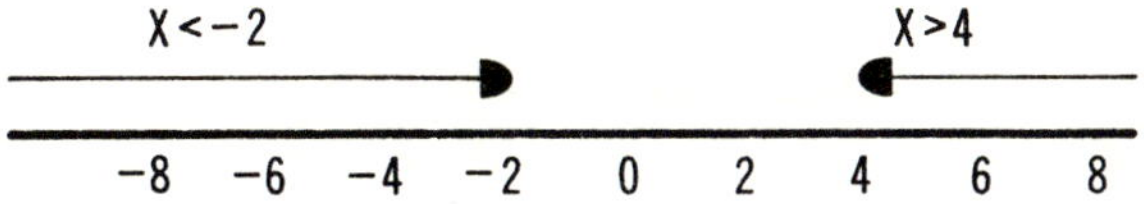

(B) Graph of [x | x > 4 or x | x < − 2].

Fig. 5-6. Graph of sets.

CHAPTER 6

The Algebra of Switching Circuits

The algebra of switching circuits is a two-valued type of Boolean algebra in which the only two possible values are 0 and 1. Suppose that 1 represents a closed circuit that allows current flow, and 0 represents an open circuit through which there is no current flow. A scheme of switching algebra can be established on this basis such that the algebra is applicable largely to series and parallel circuits. We will show that it can also be used for non series-parallel circuits.

RULES OF THE ALGEBRA

The basic circuits are the series circuit and the parallel circuit. The parallel circuit is an OR and is usually designated by switches in parallel. This may be written as A OR B, which is the same as A + B. In this algebra, the addition or + sign designates two switches in parallel. A series circuit, say of two switches, is considered to be an AND circuit. The two switches A and B in series are designated as A AND B, which is multiplication and is written A • B.

These are as defined in Table 6-1. There are three basic rules for addition, as shown. $0 + 0 = 0$ means that an open circuit in parallel with an open circuit is still an open circuit. $0 + 1 = 1$ means that an open circuit in parallel with a closed circuit is a

Table 6-1. Basic Identities of Switching Algebra

Addition	Multiplication	Negation
$0+0=0$	$0 \cdot 0=0$	$\overline{0}=1$
$0+1=1$	$0 \cdot 1=0$	$\overline{1}=0$
$1+1=1$	$1 \cdot 1=1$	
$1=$ Closed		$0=$ Open

Addition is OR, switches in parallel. Mulitplication is AND, switches in series.

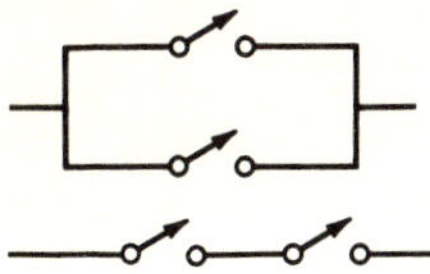

OR (A or B),(A+B)

AND (A and B),(A•B)

closed circuit. The third rule of addition is that $1 + 1 = 1$ and means that a closed circuit in parallel with a closed circuit is a closed circuit. In summary, the rules for addition say, considering a parallel circuit, there is current flow if A is closed, or B is closed, or if both are closed.

In the same way, multiplication represents a series connection. $0 \cdot 0 = 0$ means an open circuit in series with an open is still an open. $0 \cdot 1 = 0$ means that an open circuit in series with a closed circuit is still an open circuit $1 \cdot 1 = 1$ means that a closed circuit in series with a closed circuit is a closed circuit. Thus in summary for multiplication (series circuit) there is current if, and only if, all of the switches in series are closed.

The third operation for switching algebra is negation. Clearly, the negation of an open circuit is a closed circuit, and the negation of a closed circuit is an open circuit. The superior bar represents negation, as $\overline{1} = 0$, $\overline{0} = 1$.

In order to establish a valid system of switching algebra, it is necessary to establish certain rules such as those already given. There are many possibilities of establishing the basic rules or systems by which various types or forms of switching algebra could be developed. For example, 0 could represent a closed circuit, and 1 could represent an open circuit. It is also possible, since this is a two-valued system, to use any of two opposite values such as an open circuit and a closed circuit, current flow and no current flow, voltage and no voltage, high voltage and low voltage, positive voltage and negative voltage, a pulse and no pulse, or any other duals. For example, consider

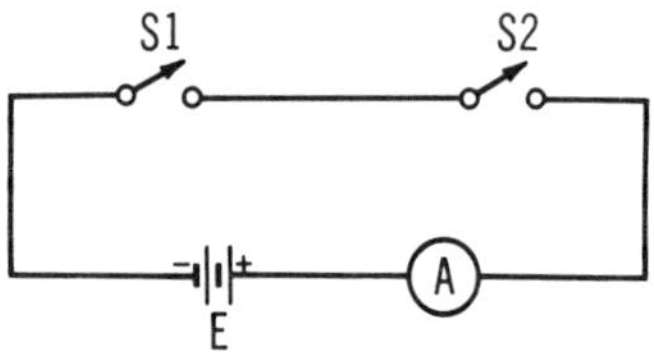

Fig. 6-1. Duality—AND/OR.

Fig. 6-1, which shows two switches S1 and S2 in series with a battery and a current-indicating device. If we were to define a 1 as meaning current flow through the meter, then there would be a 1 only if switch 1 and switch 2 were closed at the same time. Clearly, under these conditions, a 0 would exist if either switch were open. On the other hand, if a 1 is to be considered as expressing no current flow through the meter, then there will be a 1 indicated if switch S1 is open or switch S2 is open. In this manner the simple circuit shown may be either an AND circuit or an OR circuit, depending on the definitions.

There are a number of significant laws that we can use to form the axioms of this system. They are given below in an abbreviated form.

1. Commutative Laws
   a. Addition $A + B = B + A$
   b. Multiplication $A \bullet B = B \bullet A$
2. Associative Laws
   a. Addition $(A + B) + C = A + (B + C)$
   b. Multiplication $(A \bullet B) \bullet C = A \bullet (B \bullet C)$
3. Distributive Law $A \bullet (B + C) = A \bullet B + A \bullet C$
4. Identities
   a. Addition $A + 0 = 0 + A = A$
   b. Multiplication $1 \bullet (A) = A \bullet (1) = A$
5. Equality
   a. Reflexive $A = A$
   b. Transitive Where $A = B$ and $B = C$, then $A = C$
   c. Symmetric Where $A = B$ then $B = A$
6. Idempotent Laws
   a. Addition $A + A = A$
   b. Multiplication $A \bullet A = A$

The first law, which is not illustrated, is the law of closure; it states merely that addition and multiplication are always defined.

The *commutative* law has two forms. In the commutative law of addition, the order in which two quantities are added is not significant. Two quantities can be added regardless of the order of addition. In the same manner, the commutative law of multiplication says that the sequence in which two items are multiplied is not significant. These basic laws seem rather trivial; however, in structures of mathematics such as matrix theory, it is not necessarily true that multiplication obeys the commutative law.

The *associative* law is in two forms; the associative law of addition says that in adding three numbers it is not significant which two are added first. In the same manner with the associative law of multiplication, the sequence of multiplication is not significant. Another way of saying this is that when a series of numbers is added no parentheses are needed; similarly, when a series of factors is multiplied, no parentheses are needed.

The *distributive* law relates to both addition and multiplication. This law states that multiplication is distributive over addition, or the multiplier (in this case A) is distributed over both quantities in the parentheses, which are B and C.

A very significant law is the one of *identity*. The number 0 is known as the identity of addition, for, as shown, $A + 0$ or $0 + A = A$. The identity for multiplication is 1; 1 multiplied by any number A is equal to the number itself.

The law of *equality* has essentially three different aspects. In the reflexive case, the law of equality merely states that $A = A$. In the transitive case, the law says, for example, if $A = B$ and $B = C$, then $A = C$. The symmetric aspect of equality says only that if $B = A$, then $A = B$.

The *idempotent* laws are very significant to this algebra. The idempotent law for addition, for example, says that $A + A = A$. In the Boolean algebra used for switching, there is no significance to a quantity such as 2A, since $A + A = A$. Notice that this is significantly different from other forms of algebra. This is the equivalent of saying that a switch in parallel with itself is itself. The idempotent law of multiplication says that $A \cdot A = A$; this only says that a switch in series with itself is itself.

SWITCHING ALGEBRA AND CIRCUITS

There are two aspects of switching algebra; one is the algebraic representation of a given electronic switching circuit. The second is, given the algebraic expression, to form the electronic switching circuit that it represents. Both of these can be done, and the basic purpose of switching algebra is to enable the circuit designer to create specific circuits and to reduce these circuits to a minimum of complexity. For example, two circuits are shown in Fig. 6-2. One of these pictures a switch

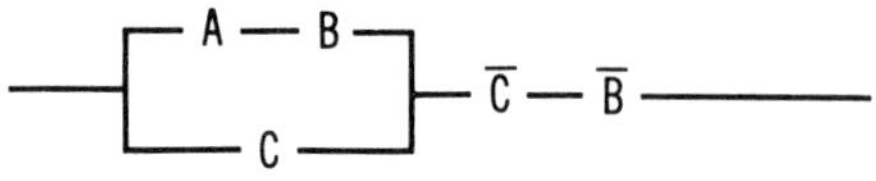

$(A \cdot B + C) \cdot (\overline{C}) \cdot (\overline{B})$

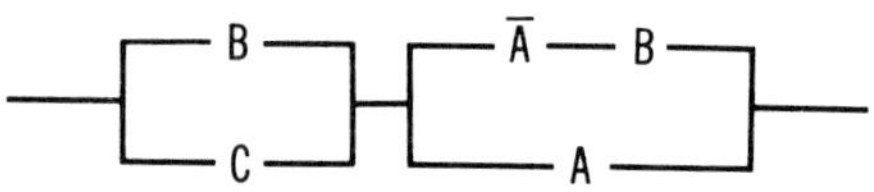

$(B + C) \cdot (\overline{A} \cdot B + A)$

Fig. 6-2. Switching circuits to represent algebraic expression.

A in series with a switch B, and both of these in parallel with switch C. This entire combination is in series with NOT C and in series with NOT B. Hence, the algebraic expression for this is:

$$(A \bullet B + C) \bullet (\overline{C}) \bullet (\overline{B})$$

The second of these is a parallel combination of switches B and C. This parallel combination is in series with a second parallel combination. The second parallel combination is a series of NOT A and B in parallel with A. Hence, the algebraic form of this is:

$$(B + C) \bullet (\overline{A} \bullet B + A)$$

One of the ways that we can evaluate a given function and its circuit is to set up a truth table such as the one shown in Table 6-2. This is shown for the AND circuit.

Consider two switches A and B, in series. These take on the four separate values shown in rows 1, 2, 3, and 4. In row 1, for

Table 6-2. Truth Table for AND Circuit

	A	B	$A \cdot B$	$\bar{A} \cdot B$	$A \cdot \bar{B}$	$\overline{(A \cdot B)}$
1	0	0	0	0	0	1
2	0	1	0	1	0	1
3	1	0	0	0	1	1
4	1	1	1	0	0	0

example, A is 0 and B is 0, so that the combination of A AND B is also 0. The next combination, which is NOT A AND B, is the equivalent of a series connection of 1 and 0, so that the result is also 0 as shown in row 1. The next combination is A AND NOT B, which has value 0. The last combination, which is the negation of A AND B, turns out to be 1. Thus it is possible, using this table, to establish the truth values for all AND combinations of switches A and B.

Truth tables are used as follows. Suppose that we have a series circuit of A AND B. Suppose also that we wish to know whether this series circuit is open or closed, if the function under consideration is the negation of A AND B. According to the table, this function is 1 except when both A and B are 1. In the same manner, if the function is A AND NOT B, the function is 1 if A is 1 and B is 0.

In the same manner, Table 6-3 shows the OR table. A OR B turns out to be 1 except when both are 0; NOT A OR B turns out to be 1 except when A is 1 and B is 0; the negation of A OR B is 1 if A is 0 and B is 0; A OR NOT B is 1 in all cases except when A is 0 and B is 1.

Table 6-4 illustrates other truth tables; each column evalu-

Table 6-3. Truth Table for OR Circuit

	A	B	$A + B$	$\bar{A} + B$	$\overline{(A + B)}$	$A + \bar{B}$
1	0	0	0	1	1	1
2	0	1	1	1	0	0
3	1	0	1	0	0	1
4	1	1	1	1	0	1

Table 6-4. All Possible Combinations of Truth Values for A and B

A	B	$\overline{A} \cdot \overline{B}$	$\overline{(A \cdot B)}$	$\overline{A} + \overline{B}$	$\overline{(A + B)}$
0	0	1	1	1	1
0	1	0	1	1	0
1	0	0	1	1	0
1	1	0	0	0	0

ates a function for all possible combinations of the truth values of A and B.

CIRCUIT SIMPLIFICATION BY TRUTH VALUES

The simplification of a circuit reduces a more complex form to a less complex form. We can always use truth tables to establish the equivalence between the original and reduced function. Consider, for example, Fig. 6-3. The original circuit is represented by:

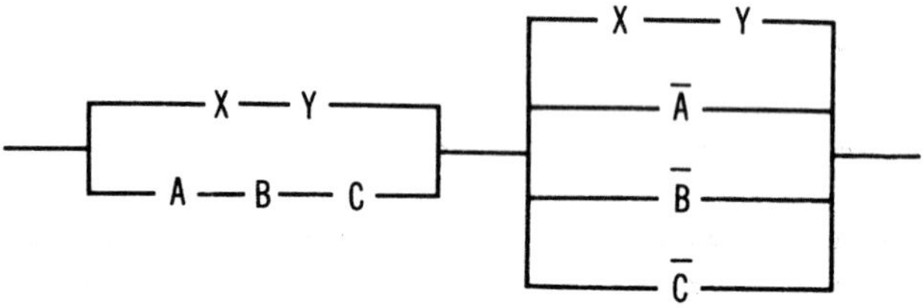

Fig. 6-3. Example of complex switching circuit.

$$(X \cdot Y + A \cdot B \cdot C) \cdot (X \cdot Y + \overline{A} + \overline{B} + \overline{C})$$

This expands to:

$$X \cdot Y + \overline{A} \cdot X \cdot Y + \overline{B} \cdot X \cdot Y + \overline{C} \cdot X \cdot Y + A \cdot B \cdot C \cdot X \cdot Y + A \cdot \overline{A} \cdot B \cdot C + A \cdot B \cdot \overline{B} \cdot C + A \cdot B \cdot C \cdot \overline{C}$$

which reduces to:

$$X \cdot Y\ (1 + \overline{A} + \overline{B} + \overline{C} + A \cdot B \cdot C) + 0 + 0 + 0$$

This, in turn, reduces to

$$X \cdot Y$$

We can also simplify expressions involving other forms. Consider this simplification:

$$X \cdot (\overline{X} + Y) + Y + Y \cdot (Y + Z) =$$
$$X \cdot \overline{X} + X \cdot Y + Y + Y + Y \cdot Y + Y \cdot Z$$

But

$$X \cdot \overline{X} = 0$$
$$Y + Y = Y$$
$$Y \cdot Y = Y$$

Thus, we have

$$X \cdot Y + Y + Y \cdot Z$$

Also,

$$Y + Y \cdot Z = Y \text{ and } Y + X \cdot Y = Y$$

And the expression becomes Y.

CIRCUIT SIMPLIFICATION BY CHARTING

A chart is a form of truth table used for simplification. Its use allows a reduction of circuit complexity.

Consider a function X of three variables, as in Fig. 6-4A, such that $X = \overline{A} \cdot B \cdot \overline{C} + \overline{A} \cdot \overline{B} \cdot C + \overline{A} \cdot B \cdot C + A \cdot B \cdot C$. This figure is a charting of the three variables where one state is 1 (closed), and the other state is 0 (open). The three variables are ordered as A, B, C, and the numbered squares are:

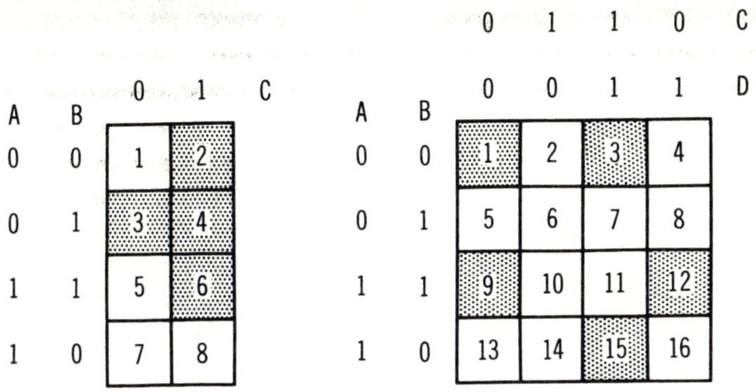

(A) Three-variable chart.

(B) Four-variable chart.

Fig. 6-4. Chart form of truth table.

1.	(0, 0, 0)
2.	(0, 0, 1)
3.	(0, 1, 0)
4.	(0, 1, 1)
5.	(1, 1, 0)
6.	(1, 1, 1)
7.	(1, 0, 0)
8.	(1, 0, 1)

The shaded blocks (2, 3, 4, 6) represent the 1 condition for the function X. If the values are (0, 1, 0) as in square 3:

$$X = 1 \bullet 1 \bullet 1 + 1 \bullet 0 \bullet 0 + 1 \bullet 1 \bullet 0 + 0 \bullet 1 \bullet 0$$
$$X = 1 + 0 + 0 + 0$$
$$X = 1$$

Again, for square 6, which is (1, 1, 1):

$$X = 0 \bullet 1 \bullet 0 + 0 \bullet 0 \bullet 1 + 0 \bullet 1 \bullet 1 + 1 \bullet 1 \bullet 1$$
$$X = 0 + 0 + 0 + 1$$
$$X = 1$$

In similar fashion, charts for five, six, seven, and even eight variables can be created. Note that a three-variable chart has eight squares, a four-variable chart has sixteen squares, and that an N-variable chart has 2^N squares. The tabular listing of the coordinate values of the variables is done in a "reflected" sequence rather than a straight binary sequence such that only one variable at a time changes state between any two adjacent coordinates. This is done to facilitate determination of redundant variables in the conjunctive terms by quick inspection.

In the same manner, in Fig. 6-4B a four-variable chart is shown where:

$$X = (A \bullet B \bullet \overline{C}) + (\overline{A} \bullet \overline{B}) \bullet (\overline{C} \bullet \overline{D} + C \bullet D) + A \bullet \overline{B} \bullet C \bullet \overline{D}$$

Here again the squares having a value of 1 are shown as 1, 3, 9, 12, and 15.

These results show for which values the function is 1, and permit a direct evaluation of the function without algebraic manipulation.

Consider Fig. 6-5, where the function charted is:

$$X = A \bullet B + \overline{B} \bullet \overline{C}$$

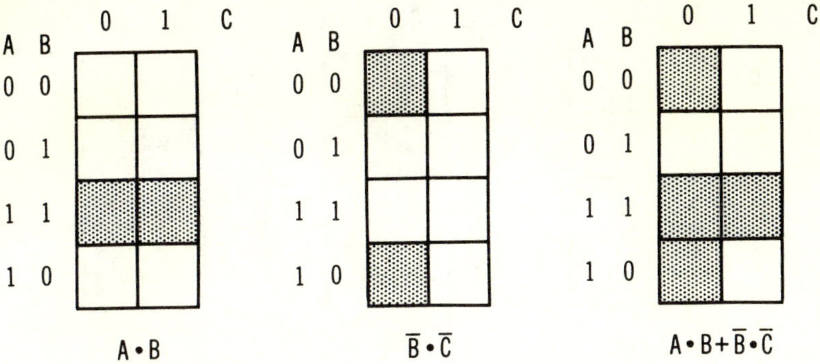

Fig. 6-5. Chart development of function $X = A \cdot B + \overline{B} \cdot \overline{C}$.

The first step is to chart $A \cdot B$ as shown. Clearly $A \cdot B = 1$ (shaded) if, and only if, $A = 1$ and $B = 1$. Values for C play no part in this, hence $A \cdot B$ is true if $C = 0$ or if $C = 1$. We can proceed to the chart for $\overline{B} \cdot \overline{C}$, which is true for $B = 0$ and $C = 0$. These two charts are then combined, as in the third chart, to produce:

$$X = A \bullet B + \overline{B} \bullet \overline{C}$$

Since this is the OR connective, the two charts are superimposed to produce the final chart.

Fig. 6-6 shows the chart development of the function $X = (A + B) \bullet (\overline{B} + C)$. Fig. 6-7 gives several charts for simple expressions. Note that $A \bullet B + \overline{A} \bullet (B + C)$ also equals $B + \overline{A} \bullet C$.

Also, formally:

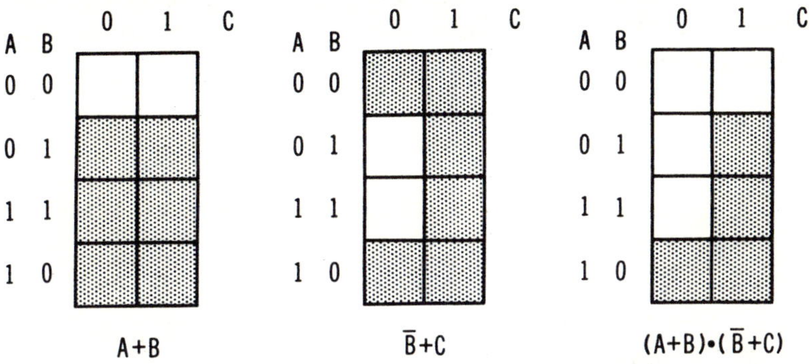

Fig. 6-6. Chart development of function $X = (A +) \bullet \overline{(B} + C)$.

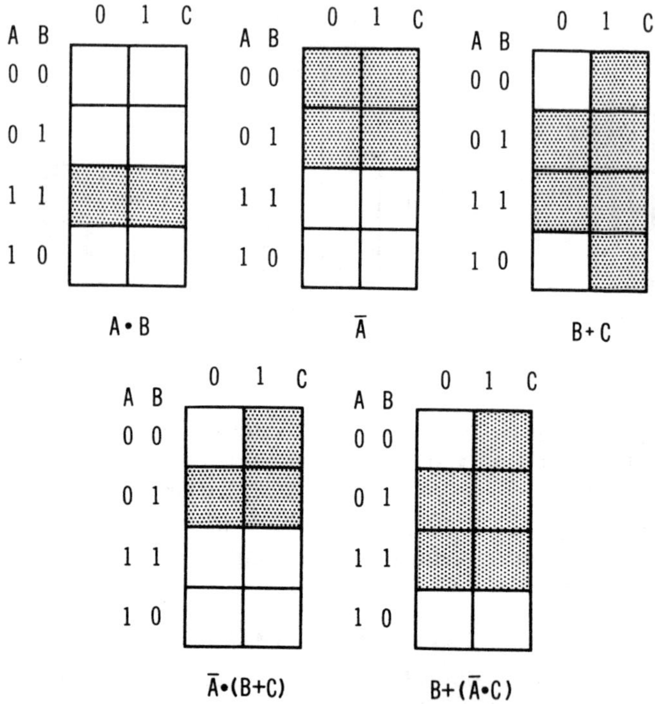

Fig. 6-7. Chart of simple expressions.

$$\begin{aligned} X &= A \cdot B + (\overline{A} \cdot B + \overline{A} \cdot C) \\ &= A \cdot B + \overline{A} \cdot B + \overline{A} \cdot C \\ &= B\ (A + A') + \overline{A} \cdot C \\ &= B + \overline{A} \cdot C \end{aligned}$$

Consider now the three-variable chart (Fig. 6-4A). Each square is written as an N-legged AND gate, where N is the number of variables.

Square	Gate
2	$\overline{A} \cdot \overline{B} \cdot C$
3	$\overline{A} \cdot B \cdot \overline{C}$
4	$\overline{A} \cdot B \cdot C$
6	$A \cdot B \cdot C$

The shaded squares indicate the values of A, B, and C for which the value of the function $X = \overline{A} \cdot \overline{B} \cdot C + \overline{A} \cdot B \cdot \overline{C} + \overline{A} \cdot B \cdot C + A \cdot B \cdot C$ is 1. The unshaded squares indicate the values for A, B, and C for which the function X is 0. Each shaded square defines a configuration of a three-legged AND gate; this is given by those values of A, B, and C, which, formed in an AND combination, have a value of 1. For example, square 2 of Fig. 6-4A is shaded, and the values of A, B, and C that define it are $A = 0$, $B = 0$, and $C = 1$. Hence, a combination of values for an AND gate (representing square 2) is $\overline{A} \bullet \overline{B} \bullet C$, since $1 \bullet 1 \bullet 1 = 1$. Thus, square 2 represents an AND gate that is high for an input of $\overline{A} \bullet \overline{B} \bullet C$, where $\overline{A}$ is a high input, and $\overline{B}$ is a high input, and C is a high input. Note the gates represented by squares 3, 4, and 6. Forming all of the gates in an OR combination, we construct a circuit whose functional expression is X. This is because the function X has a value of 1 whenever *any* of these gates has a value of 1.

In a four-variable case (Fig. 6-4B) the expression could be written as:

$$\overline{A} \bullet \overline{B} \bullet \overline{C} \bullet \overline{D} + \overline{A} \bullet \overline{B} \bullet C \bullet D + A \bullet B \bullet \overline{C} \bullet \overline{D} + A \bullet B \bullet \overline{C} \bullet D + A \bullet \overline{B} \bullet C \bullet D$$

This is obtained by considering the squares 1, 3, 9, 12, and 15 in sequence, each as a four-legged AND. This is also expressed in the form of the original function:

$$A \bullet B \bullet \overline{C} + (\overline{A} \bullet \overline{B}) \bullet (\overline{C} \bullet \overline{D} + C \bullet D) + A \bullet \overline{B} \bullet C \bullet D$$

as shown in Fig. 6-8A.

The simplification of the expression is achieved by inspecting the chart for shaded large squares or rectangles composed of the individual shaded squares such that all of the individual squares are shaded and such that the number of individual squares composing the larger shaded square or rectangle is one, two, four, or eight. *The two outside columns of the chart are considered to be contiguous* for purposes of this inspection. Similarly, *the top and bottom rows of the chart are also considered to be contiguous to each other.*

There are rules for simplification that are often used. Here are several:

1. Four adjacent squares (a 2×2 rectangle or 4×1 rectangle) form a two-legged gate.
2. Two adjacent squares form a three-legged gate.
3. Single squares form a four-legged gate.

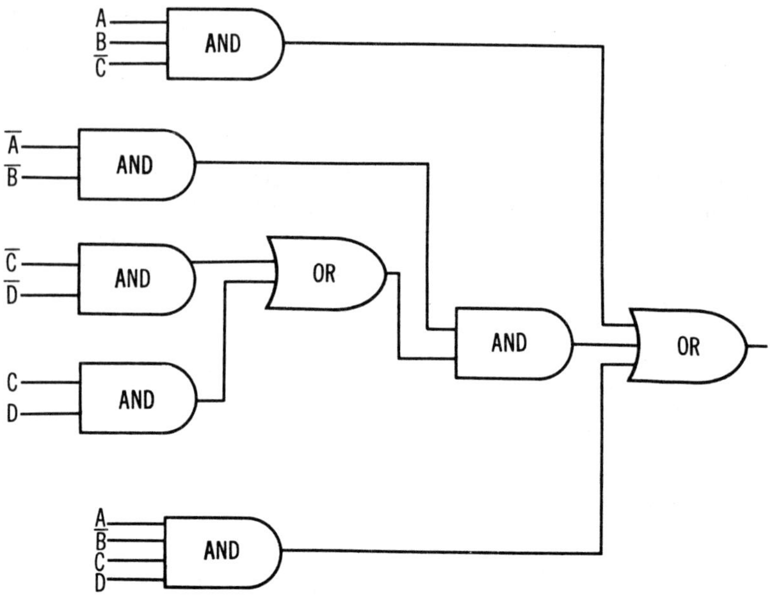

(A) Original circuit.

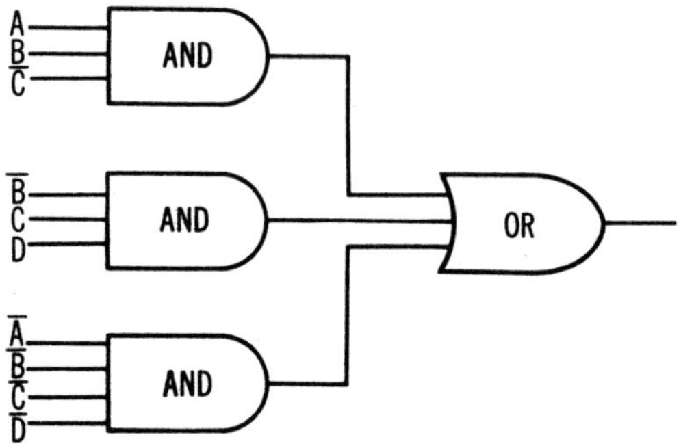

(B) Simplified circuit.

Fig. 6-8. Simplifying circuits by reducing algebraic expression.

With the four-variable example, in Fig. 6-4B:

Squares	Representation
9, 12	$A \bullet B \bullet C$
3, 15	$\overline{B} \bullet C \bullet D$
1	$\overline{A} \bullet \overline{B} \bullet \overline{C} \bullet \overline{D}$

Hence the reduced function is:

$$A \cdot B \cdot \overline{C} + \overline{B} \cdot C \cdot D + \overline{A} \cdot \overline{B} \cdot \overline{C} \cdot \overline{D}$$

This is shown in Fig. 6-8B.

For the three-variable case (Fig. 6-4A), the function is:

$$\overline{A} \cdot B \cdot \overline{C} + \overline{A} \cdot \overline{B} \cdot C + \overline{A} \cdot B \cdot C + A \cdot B \cdot C$$

Squares 3 and 4 give $\overline{A} \cdot B$; squares 2 and 4 give $\overline{A} \cdot C$; squares 6 and 4 give $B \cdot C$. This reduces to:

$$\overline{A} \cdot B + \overline{A} \cdot C + B \cdot C$$

CHAPTER 7

Numbers and Numbering Systems

Switching-circuit algebra rapidly leads to an examination of the numbering system that is used with this algebra. The switching circuits that have been discussed can be used as the basis of logical design in various types of control configurations; these switching circuits can be arranged in logical systems to perform certain necessary switching functions. These circuits work on a go–no-go basis, which means that they either pass a signal, or they do not pass a signal. In this sense, circuits do not handle numbers as such.

Switching circuits, however, can be designed to perform arithmetic functions such as addition, subtraction, multiplication, or division. In order to do this, a numbering system must be defined. This is quite different from the use of the switching circuits in logical design. In arithmetic circuits we are concerned with the manipulation of various number representations. However, such arithmetic calculating circuits are based upon the fundamentals of algebraic switching and of logical design.

BINARY NUMBERS

There are many possibilities for numbering systems; the one that is in common use in our civilization is the decimal notation based upon powers of ten. For example, the number 732 means seven hundred thirty-two. This number also means seven times one hundred plus three times ten plus two times one. The same number can be expressed as seven times ten squared plus three times ten to the first power plus two times ten to the zero power (since any number to the zero power is one).

The base of our decimal system is, of course, ten, since in moving from right to left in any number we increase by one power of ten for each digit that we move to the left. In a similar manner, by moving to the right we decrease by one power of ten.

There are many possibilities for numbering systems where other bases can be used. One of the most convenient numbering bases is two. This is the binary numbering system, in which any digit can have only one of two possible values; one of these values is 0 and the other is 1. In the decimal numbering system, of course, it is possible to have ten different indications which are 0, 1, 2, 3, 4, 5, 6, 7, 8, 9.

The binary system corresponds well to the requirements of circuits designed by the application of Boolean algebra, since Boolean algebra is a two-valued system. It is possible, for example, in a truth table to have a proposition be true or false, and in the same way it is possible to give values such as 0 or 1. Usually 1 corresponds to true, and 0 corresponds to false. The binary number system may be used for calculating circuits using binary arithmetic.

The first five powers of ten are 1, 10, 100, 1,000, and 100,000. The first five powers of 2 are 1, 2, 4, 8, and 16. The decimal numbers 1, 2, 3, 4, 5, and 6 correspond to the binary numbers 1, 10, 11, 100, 101, and 110. This means, for example, that the binary number 101 is the equivalent of 4 plus 0 plus 1, which is decimal 5 (Table 7-1).

The binary numbering system makes it extremely convenient to use Boolean algebra since only one of two possible states has to be considered. These states are the 1 state and the 0 state. For example, consider a vacuum tube or a transistor as the circuit element representing a digit. If the transistor is on, this represents a 1; if the transistor is off, this represents a 0. In the same way, if the vacuum tube is conducting current, this could be a 1, and if the vacuum tube were not conducting, this would be a 0. The binary system is the most convenient num-

Table 7-1. Decimal to Binary Conversion

Trigger	Readout 2^N	2^{N-1}	. . . 2^1	2^0
Reset	0	0	0	0
1st	0	0	0	1
2nd	0	0	1	0
3rd	0	0	1	1
4th	0	0	0	0
5th	0	0	0	1
6th	0	0	1	0
7th	0	0	1	1
•	•	•	•	•
•	•	•	•	•
•	•	•	•	•
N — 1	1	1	1	1
N	0 (Carry)	0	0	0

bering system, since the active circuit elements such as transistors, vacuum tubes, diodes, or various types of magnetics, have two positive states (either on or off); thus there is no ambiguity. The use of a decimal numbering system, for example, would require an active circuit element with ten clearly distinct states. This is not impossible, but it is much more difficult to achieve than circuits using the binary numbering system.

BINARY ARITHMETIC

We must study the manipulation of binary numbers before it is proper to examine the circuit arrangements by which the logical design manipulates the numbers. For example, consider addition, the fundamental arithmetic operation. The concept of addition implies that there are two numbers; one is called the augend, and the other is called the addend. When these two are added, the result is the sum. This is obvious in decimal arithmetic where, for example, the sum of 8 and 9 is 17. Note, however, even in this simple example there are now two digits in the sum, though there was only one digit in each of the two numbers that were added.

For binary addition there are only the two digits 0 and 1. There are four possible cases. These are:

$$0 + 0 = 0$$
$$1 + 1 = 0$$
$$0 + 1 = 1$$
$$1 + 1 = 10$$

The only one of these above that may create a problem is that $1 + 1 = 10$. Suppose, for example, that you have a decimal dial counter such as an automobile odometer. The number on the right begins at 0, and goes from 1 through 9. After you have gone more than 9 miles you have exceeded the range of the right hand digit, and the counter turns so that the rightmost digit becomes 0 and the left digit becomes 1. In the same manner the $1 + 1 = 10$ indicates that you have exceeded the range of values of the rightmost digit in the binary system.

Actually, all of the other arithmetic operations are defined in terms of the operation known as addition. For example, subtraction is negative addition, multiplication is just a shorthand way of adding a long series of numbers, and division is the inverse of multiplication.

CODED NUMBERING SYSTEMS

The binary numbering system that we have discussed is known as a pure binary system, where each individual bit (binary digit) has a weighted decimal value. There is a direct conversion between this binary system and its equivalent decimal. For example, the binary numbers 001, 010, and 011 are the equivalent of the decimals 1, 2, and 3. There are limitations to the use of pure or straight binary numbers, and for this reason codes are often used.

One of these is the *binary coded decimal,* which is a special way of expressing decimal numbers in terms of binary numbers. For example, consider number 18 in decimal. If this decimal is expressed in pure binary, it is 10010. Evaluating this binary number, place for place, gives a series of weighted values that add up to the decimal number 18. This binary number is the equivalent of decimal 16 plus decimal 2, or decimal 18.

However, in the binary coded system a different technique is used. If the two digits in the number 18 are evaluated separately, the number 1 is evaluated and its expression in binary coding is 0001. The 8 is evaluated separately and its coding is

1000. In this way we can convert individual digits such as 1 or 8, using the binary coded number system rather than converting a group of digits such as 18. This system is used because conversion from decimal to binary, or from binary to decimal, is not convenient. The conversion from binary coded decimal the decimal unit from which it is converted. For example, the straightforward, and convenient. In this way, the binary coded decimal system allows us to use binary numbers to code any decimal number on a digit-by-digit basis. Each of the individual digits is given a binary coding.

One of the other codes that is often used is the *excess three*. In this type of coding each binary number is represented by a method in which the binary number is three units greater than the decimal unit from which it is converted. For example, the decimal 1 appears as the straight-binary equivalent of the decimal 4; the decimal 3 appears as the straight-binary equivalent of the decimal 6. The reason the excess-three code is used is that it provides a simplification of certain types of arithmetic so that complements may be used directly.

Another group of codes are the *reflected cyclic codes*. These codes have one thing in common; in going from one coded decimal digit to the next coded decimal digit only one of the coded digits changes at a time. For example, in straight, or pure, binary, a decimal 2 is 0010, and a decimal 3 is 0011. Note that there is only one digit changed in going from 0010 to 0011. However, in going from decimal 7 to decimal 8 there is a change from 0111 to 1000, a change in *four* digits. In the *gray* code, which is one type of reflected cyclic code, the change from decimal 7 to decimal 8 involves a change of only one binary digit, i.e., 0100 to 1100. The gray code is used because the change of only a single digit in going from one decimal value to another is convenient in analog-to-digital conversion.

USING COUNTING CIRCUITS

The preceding sections in this chapter have outlined the basic principles of binary numbers and the binary number system. These numbers, of course, may be used for calculating circuits in which addition, subtraction, multiplication, or division is possible. This section covers the manipulation of these numbers by counting circuits.

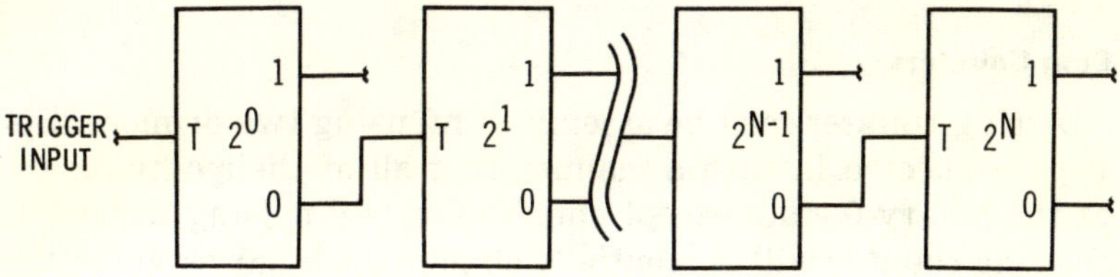

Fig. 7-1. Binary counter.

The circuit element that is used for a counter is a binary flip-flop. This is the electrical equivalent of a toggle switch having two positions. Just as a toggle switch is either on or off, so a binary flip-flop has two states; these are the 1 and 0 state. Thus a binary flip-flop is a counter, and it counts in sequence such as 0, 1, 0, 1, etc. A series of flip-flops may be used for the counting in the binary numbering scheme.

Binary Counters

A straight binary counter may be assembled by using one or more flip-flops connected in such a manner that the binary number stored within these flip-flops will represent the total number of trigger pulses received at the input to the counter (Fig. 7-1).

This (Fig. 7-1) is a series of flip-flops, each representing a power of 2. To determine a readout of a flip-flop, the binary state of the 1 terminal (the terminal adjacent to the set side of the flip-flop) is used as the output for a positive-logic binary-state indicator; the 0 terminal may be used to drive a negative-logic binary indicator. If a positive-logic binary-state indicator is connected to the flip-flop output, the indicator presents the indications given in Table 7-2. A binary 1 is equivalent to the "on" condition of the element.

Table 7-2. Binary Counter Indications

Binary	Decimal
0 0 0	0
0 0 1	1
0 1 0	2
0 1 1	3
1 0 0	4
1 0 1	5
1 1 0	6
1 1 1	7

Ring Counters

A ring counter may be assembled by using two or more flip-flops connected in such a manner that all of their outputs are at the binary 0 state except one flip-flop. By pulsing the input, the ring counter will sequentially change the binary state of the succeeding flip-flop from a binary 0 to a binary 1. The flip-flop that contains the binary 1 indicates the count of the counter. The number of pulses that can be counted by N flip-flops is N pulses.

Shift Registers

A serial-entry shift register is similar to the ring counter with the exception that the output flip-flop is not connected to the input flip-flop. The serial binary information is applied to the first flip-flop set and reset gates. All set-trigger and reset-trigger inputs are tied together to form the *shift bus.* Clock pulses are applied to the shift bus to cause the binary information to shift from left to right, one bit position for each clock pulse received.

Preset Counters

A preset counter may be assembled by using two or more flip-flops connected as a straight binary counter or as a feedback counter (binary coded decimal, excess three, etc.). A number recognition gate (AND gate) is connected to the output of each individual flip-flop in the counter. When the desired number has been reached by the counter, an AND gate, through an inverter, or a NAND gate, inhibits the input to the counter and prevents the progression of the counter beyond the desired number. The output from this AND gate may also be used to provide an output indication that the preset counter has reached the desired number.

A preset counter may also be assembled by using two or more flip-flops connected as a backward counter. A straight binary backward counter with an AND gate to recognize the number zero is shown in Fig. 7-2. The counter is originally preset to the desired number. This can be accomplished by the set and reset trigger inputs or by the dc set and reset inputs. Incoming trigger pulses cause the counter to count backwards; that is, each trigger pulse subtracts one count from the counter. This proc-

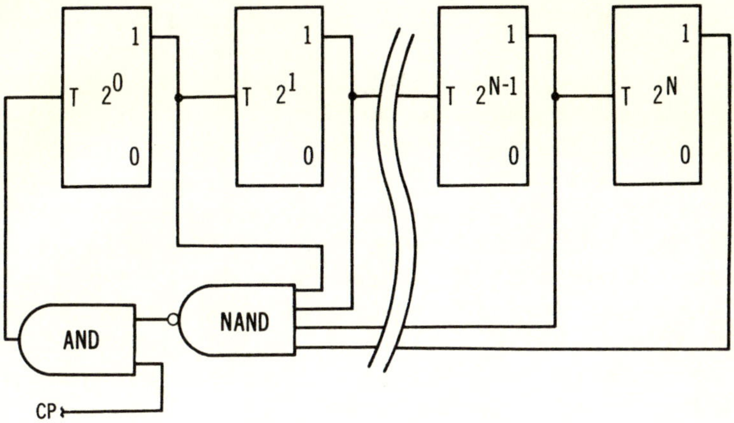

Fig. 7-2. Binary backward counter.

ess would continue through zero and repeat in an endless cycle starting with 2^{N-1} (provided no feedback is used). Therefore, an AND gate is used to recognize the unique count of zero and inhibit the input to the counter. The output from this AND gate can also be used to provide the external signal to indicate that the preset counter has reached the desired number (received the desired number of input pulses).

A preset counter connected as a backward counter with zero detection may be used to generate a precise time delay. The counter starts subtracting counts from its preset number immediately after a preset condition, if the trigger pulses are in a continuous pulse train, e.g., clock pulses.

Feedback Counters

Feedback counters may be assembled by using two or more flip-flops connected so as to recycle when a specific number has been reached. There are many different ways of providing a recycle count at a desired number. One method that always works to provide a recycle count at a desired number is to recognize $N - 1$ counts with a number recognition gate (AND gate), and cause the counter to reset on the next incoming trigger pulse. A more sophisticated approach is to use feedback to cause recognition at a desired count. Care must be exercised in using feedback if a coded binary readout is desired. Another method that may be employed is to use combinations of known counter connections to recycle at the desired number. In such a configuration, to maintain a binary coded readout, only straight binary counters may be placed in front of the feedback counter, and only in *front* of the feedback counter; that is, the straight binary counter may not be placed after the feedback counter. An example of this method is shown by an N/12 counter, where an N/4 and an N/3 counter are employed.

An N/3 counter and its truth table are shown in Fig. 7-3. This is a binary-coded ternary counter. The output of the counter is returned to the set gate of the first flip-flop in order to inhibit this flip-flop for every third incoming pulse. In addition, the complement of the counter output is returned to the reset gate of the second flip-flop in order to permit the second flip-flop to reset every third pulse.

An N/7 counter and its truth table are shown in Fig. 7-4. This counter presents an excellent example of "don't care" conditions. There are certain bistable states, such as the condition of the counter after every sixth pulse, in which only the last two flip-flops are of interest. Therefore, all that is necessary is to recognize the condition of a binary 1 in the last two flip-flops and to cause a reset to occur with the next trigger pulse.

Comparators

If it is desired to determine if two serial binary numbers are identical, an equalizer circuit, using the logic in Fig. 7-5 may be employed. This logic may be simplified by using an exclusive OR and an inverter as shown in Fig. 7-6. If it is required to compare two serial binary numbers to determine which number is

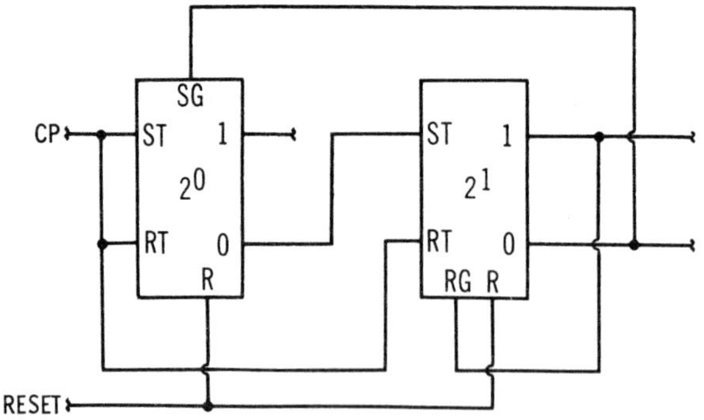

	2^0	2^1
RESET	0	0
1	1	0
2	0	1
3	0	0
4	1	0
5	0	1
6	0	0

Fig. 7-3. N/3 counter and truth table.

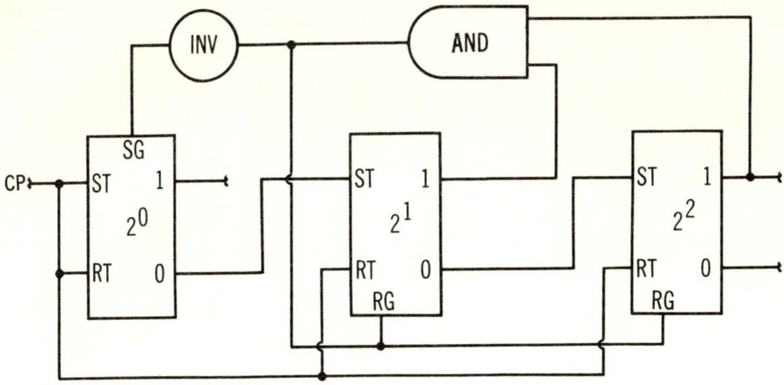

	2^0	2^1	2^2
RESET	0	0	0
1	1	0	0
2	0	1	0
3	1	1	0
4	0	0	1
5	1	0	1
6	0	1	1
7	0	0	0

Fig. 7-4. N/7 counter and truth table.

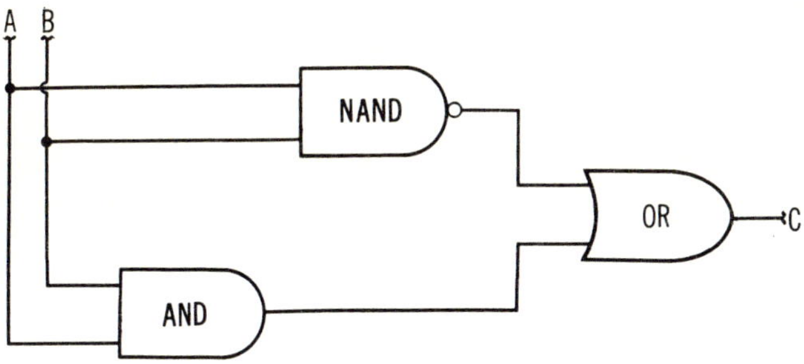

Fig. 7-5. Equalizer circuit logic.

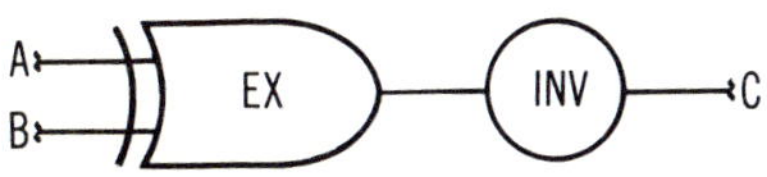

A	B	C
1	1	1
1	0	0
0	1	0
0	0	1

Fig. 7-6. Simplified circuit for logic used in Fig. 7-5.

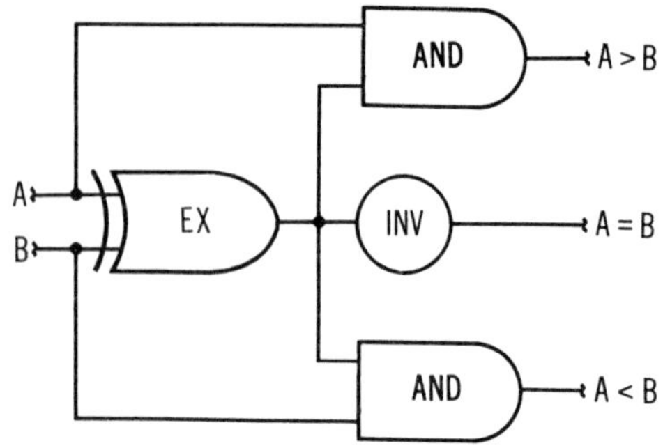

INPUT		OUTPUT		
A	B	A > B	A = B	A < B
1	1	0	1	0
1	0	1	0	0
0	1	0	0	1
0	0	0	1	0

Fig. 7-7. Comparator logic.

larger, or if it is desired to compare an incoming serial binary number with a reference number in order to determine whether the incoming number is greater than, less than, or equal to the reference number, the logic in Fig. 7-7 may be used.

Binary to Decimal Conversion

There are many methods of converting a binary number to a decimal number. Some of these methods are:

1. The division-by-10 process, with the remainder after each division operation indicating the correct decimal digit.
2. The subtraction-of-10 process.
3. Direct matrix conversion.
4. The use of forward-backward counters.

However, if the binary number assumes an appreciable magnitude, then the equipment involved may become cumbersome, or the time may not be available to perform the conversion.

In one system, the decimal value of a binary number may be determined by adding the powers of 2, as indicated by the binary weighting (2^0, 2^1, 2^2, etc.) of the binary number. Convenient use is made of the fact that when a binary number is shifted one bit toward the most significant digit in a binary register, it is equivalent to multiplying the binary number by 2. The shift register may be divided into groups of four flip-flops, called *decades*. The first decade carries a decimal weighting of one; the second a decimal weighting of ten; the third, a weighting of one hundred, etc. If a serial binary number of N bits is shifted into the register with the most significant digit first, the second bit will be doubled N – 1 times and the following bits by N – 2 times, N – 3 times, etc.

However, in shifting each of these bits from one decade (with a decimal value of 8) to the next decade (with a decimal value of 10) there is an error of 6. To correct this error, an adjustment process must be made prior to each shift. The condition of each decade should be sensed; and, if the value of a decade is 5 or greater, a value of 3 must be added to that decade prior to the shift. The simplest process by which 3 may be added to each decade is to use a complementing process as shown below.

Binary/BCD (+3)	**Complement**	**BCD/Binary (−3)**
5	8-4-1	8
6	8-4-2-1	9
7	8-4-1	10
8	2-1	11
9	4-1	12

Since there is no propagation of a carry to a higher decade by this method, each decade may be treated individually, and as many decades may be cascaded as desired without any interconnecting logic. One decade (4 flip-flops) is required for each decimal digit desired.

CHAPTER **8**

Applications of Switching Circuits

The preceding chapter covered the numbering system used in most logical calculating circuits. This chapter includes switching circuits, and their application to calculating circuits.

NON SERIES-PARALLEL CIRCUITS

There are limitations to the use of Boolean algebra in the design and analysis of switching circuits. In general, Boolean algebra is a convenient tool for the analysis of series and parallel circuits. There are essentially no problems in representing a series or a parallel circuit by Boolean algebra expressions, or, conversely, of representing a Boolean algebra expression by a series or parallel or combination circuit.

There are, however, major problems in treating non series-parallel circuits with the Boolean algebra techniques. Yet, it is possible by the techniques shown in this section to convert many non series-parallel circuits into their Boolean algebra form. In general, however, it is not possible to take the Boolean algebra expression for a non series-parallel circuit and from this expression derive the actual non series-parallel electronic circuit that this represents. Only in special cases that are discussed later is there a satisfactory technique for translating

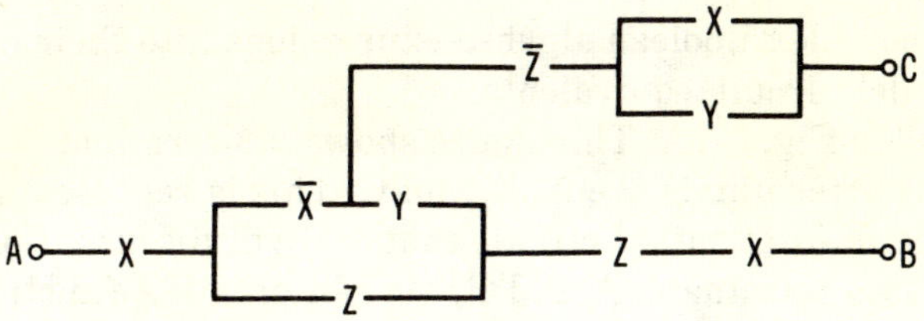

$$f_1 = f_{AB}$$
$$f_2 = f_{AC}$$
$$f_3 = f_{BC}$$

(A) Three-terminal network.

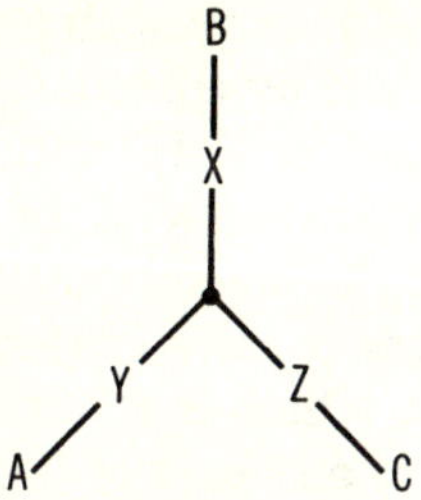

(B) Wye.

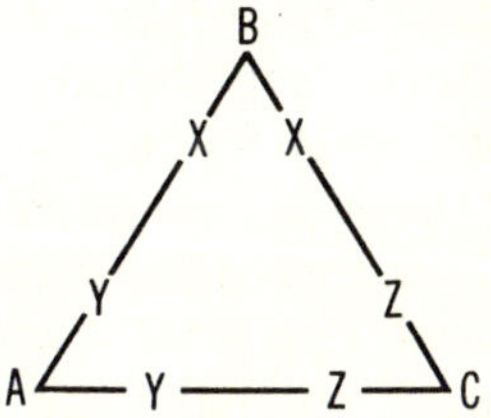

(C) Delta.

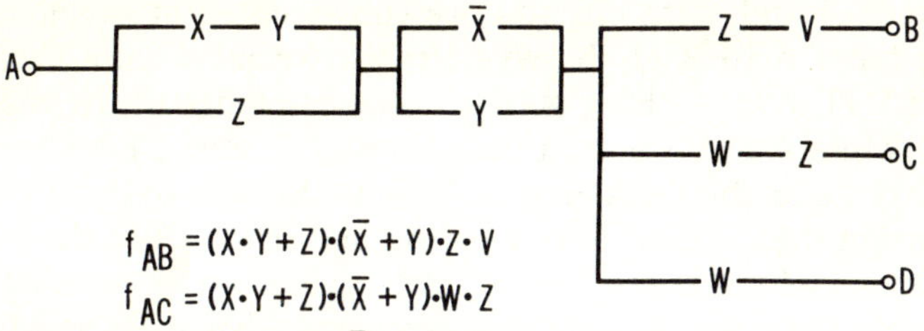

$$f_{AB} = (X \cdot Y + Z) \cdot (\overline{X} + Y) \cdot Z \cdot V$$
$$f_{AC} = (X \cdot Y + Z) \cdot (\overline{X} + Y) \cdot W \cdot Z$$
$$f_{AD} = (X \cdot Y + Z) \cdot (\overline{X} + Y) \cdot W$$

(D) Four-terminal network.

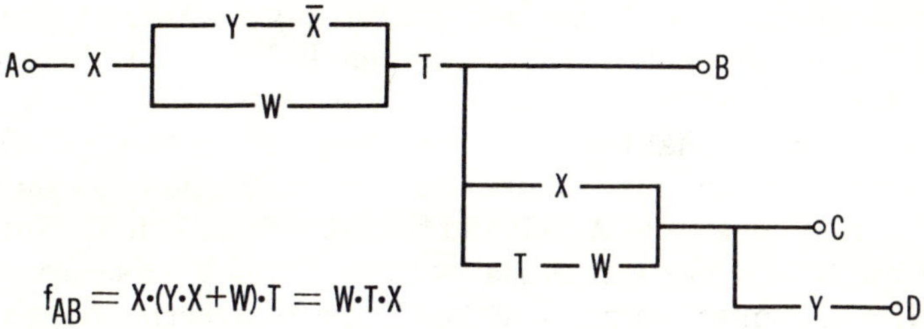

$$f_{AB} = X \cdot (Y \cdot X + W) \cdot T = W \cdot T \cdot X$$
$$f_{AC} = X \cdot (Y \cdot X + W) \cdot T \cdot (X + T \cdot W) = W \cdot T \cdot X$$
$$f_{AD} = X \cdot (Y \cdot X + W) \cdot T \cdot (X + T \cdot W) \cdot Y = X \cdot W \cdot T \cdot Y \cdot (X + T \cdot W) = X \cdot W \cdot T \cdot Y$$

(E) Variation of four-terminal network.

Fig. 8-1. Treatment of non series-parallel circuits.

certain complex Boolean algebra expressions into their non series-parallel circuit equivalents.

Consider Fig. 8-1A. This figure shows a 3-terminal network; these three terminals are A, B, and C. This is representative of the basic multiterminal circuit that we will discuss. Considering the two terminals, A and B, we can obtain a function that relates the switching elements between these two terminals. In the same terms we can obtain a second function relating the switching circuits between A and C. The third function represents the switching circuits between terminals B and C.

We can generalize the problem in transforming such a multiterminal circuit into a series-parallel circuit by considering the transformation also shown in this figure. Fig. 8-1B is the wye circuit, again having three terminals and three switches. However, there is a common central point in the wye circuit where the three 2-terminal circuits have a common point instead of a terminal. Fig. 8-1C shows a delta circuit, which is a 3-terminal circuit in which the only common points of any pair of the three 2-terminal circuits are three terminals A, B, and C. Thus the wye circuit and delta circuits are equivalent. For example, in going from A to B of the wye circuit, we must pass through X and Y. The same thing is true in the delta circuit. In passing from points A to C we must pass through Y and Z in either the wye circuit or the delta circuit. This is the wye to delta transformation that is used in converting one circuit into the other, and is one of the bases for translating non series-parallel circuits into their series-parallel equivalents. Figs. 8-1D and E are other multiterminal circuits.

An extension of this transformation is the star-to-mesh transformation, which can be used for any number of terminals. Fig. 8-2 shows this transformation. In Fig. 8-2A, there are four terminals shown and four switches, which are W, X, Y, and Z. Note also that there is a common point. In this transformation the first step is to redraw the two connections, as shown in Fig. 8-2B, one from A to B and the other from B to C. Notice that the A to B connection has switches X and Y in series, and the B to C connection has switches Y and W in series. This can be seen from Fig. 8-2A.

Two of the remaining legs are shown in Fig. 8-2C. These are the legs A to D and C to D. In Fig. 8-2D, the entire transformation is shown, including the two additional legs from B to D

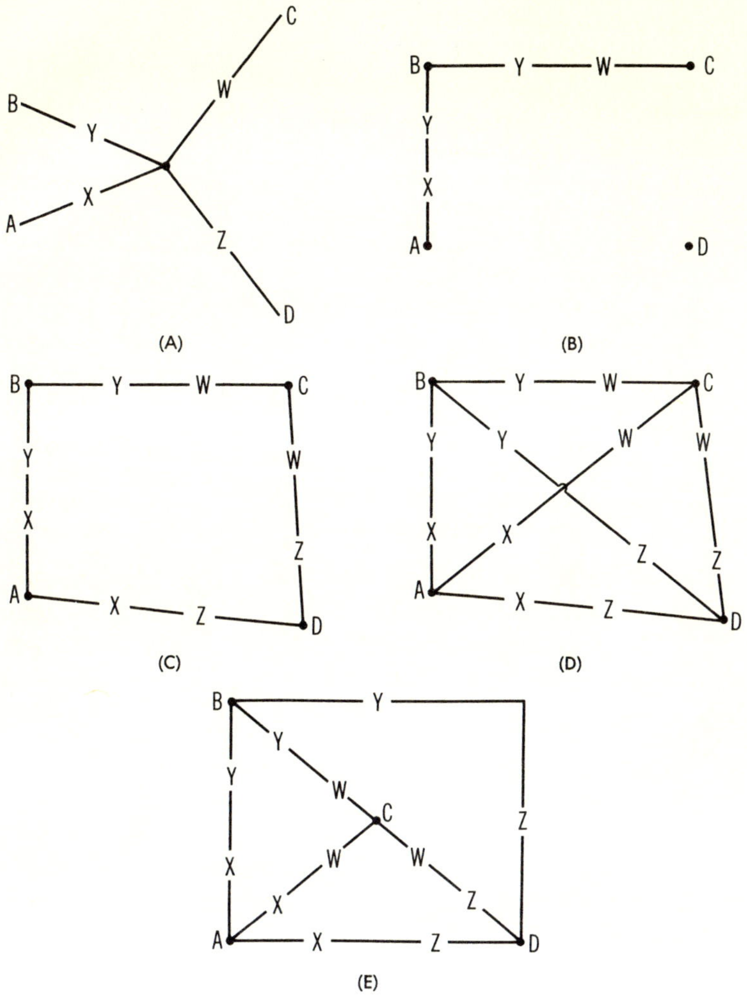

Fig. 8-2. Star-to-mesh transformation.

and from A to C. This same transformation can be redrawn as in Fig. 8-2E. The common central point has now been eliminated. It is, in general, possible to apply the star-to-mesh transformation in order to convert a circuit into one that may be treated by Boolean algebra.

Consider the bridge circuit shown in Fig. 8-3. This is a 2-terminal non series-parallel circuit. There are four possibilities by which a current flow could be completed from A to B. These

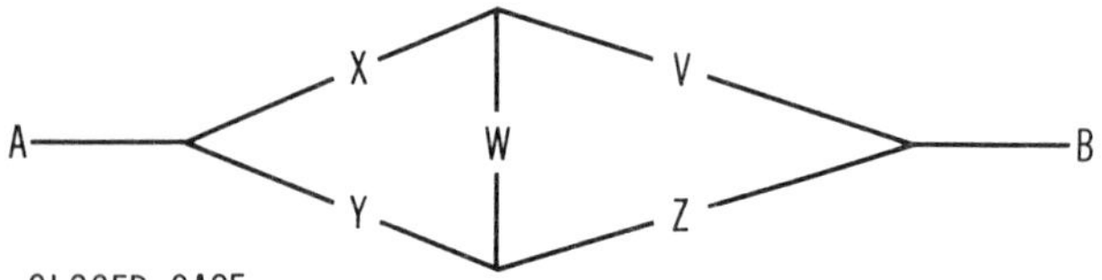

CLOSED CASE

$f_1 = X \cdot V$
$f_2 = Y \cdot Z$
$f_3 = X \cdot W \cdot Z$
$f_4 = Y \cdot W \cdot Z$

$\therefore f_{AB} = f_1 + f_2 + f_3 + f_4$

$f_{AB} = X \cdot V + Y \cdot Z + X \cdot W \cdot Z + Y \cdot W \cdot Z$

OPEN CASE

$f_1 = X + Y$
$f_2 = V + Z$
$f_3 = X + W + Z$
$f_4 = Y + W + V$

$\therefore f_{AB} = (f_1)(f_2)(f_3)(f_4)$

$f_{AB} = (X+Y) \cdot (V+Z) \cdot (X+W+Z) \cdot (Y+W+V)$

Fig. 8-3. Two-terminal non series-parallel circuit.

four possibilities are shown as four functions in the figure. For example, if switch X and switch V were closed, there would be a complete path from A to B. Also, if switches Y and Z were closed, there would be a complete path. In the same manner a complete path would exist if switches X, W, and Z were closed, or if switches Y, W, and V were closed. Thus the complete function (AB) can be considered as four possible paths.

Another alternative is to consider the possible ways in which this circuit could be opened. If switches X and Y are both open, there is no current flow. In the same manner, if switches V and Z are both open, there is no current flow. Two similar methods for preventing current flow are to open the switches X, W, and Z or to open the switches Y, W, and V. Thus, there are four possible ways of opening the circuit. In this manner, it is possible to write a different expression for the same function. Based on this, we can take the bridge circuit shown in Fig. 8-4A and redraw it as any of the three circuits shown in Figs. 8-4B, C, or D. All four of the circuits in Fig. 8-4 are electrically equivalent.

SYMMETRIC FUNCTIONS

There is a special type of Boolean function that represents the *symmetric* circuit. This is a type of non series-parallel cir-

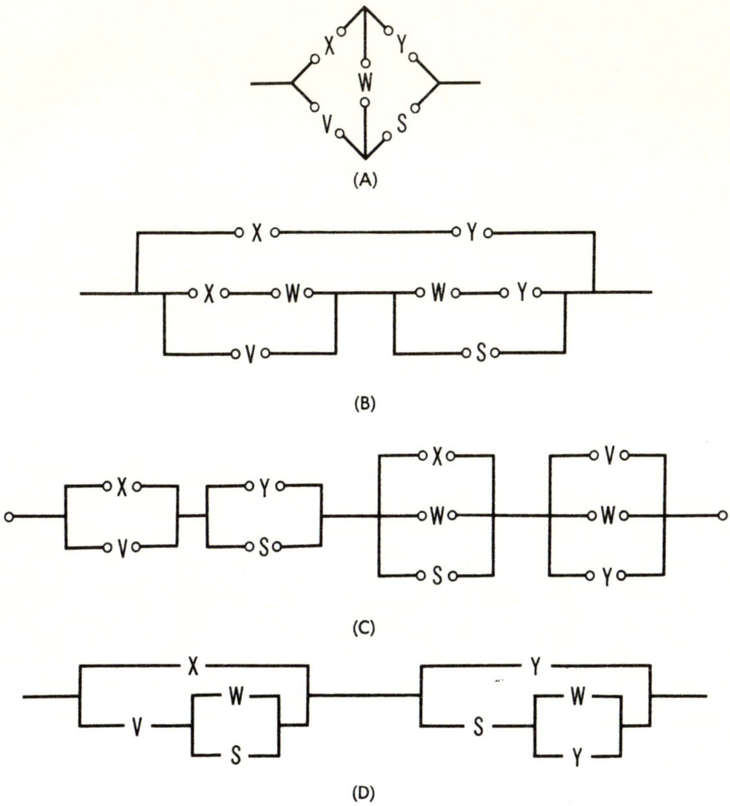

Fig. 8-4. Electrically equivalent bridge circuits.

cuit that often occurs and for which established methods of solution are known. A function is said to be symmetric if, and only if, the interchange of any pair of variables leaves the function unchanged.

Consider the three-function symmetric circuit shown in Fig. 8-5. There are five terminals; A, B, C, D, and E. There are three functions represented by both the functions and their complements. Note that in this figure it is possible to interchange any pair of variables and obtain exactly the same function. In order to have a closed circuit from A to C there are several possible paths. Each of the junctions, 1, 2, 3, 4, and 5, is numbered for convenience. For example, in going from point C to point A, we can have the three closed switches; X, Y, and $\overline{Z}$. It is also possible to go from C to A by means of $\overline{X}$, Y, and Z. Another

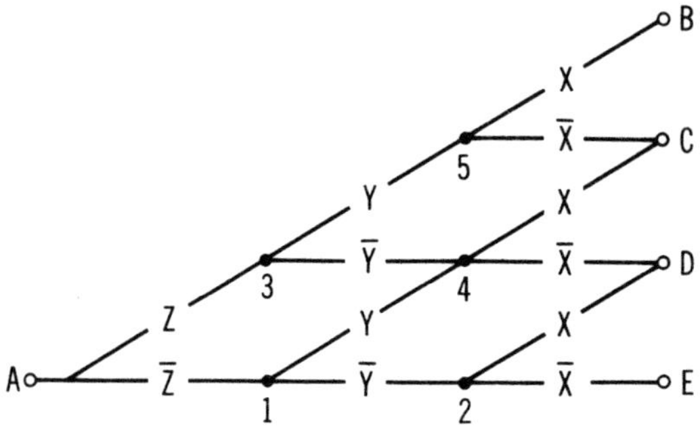

Fig. 8-5. Circuit representing three symmetric functions.

possible path is X, $\overline{Y}$, and Z. Note that we cannot go through the path $\overline{X}$, Y, $\overline{Y}$, Y, and $\overline{Z}$ since Y and $\overline{Y}$ cannot both be closed at the same time. In order to go from C to A in this symmetric function of three variables, it is necessary and sufficient to close only two switches.

In tracing the path from C to A there are several possibilities.Consider first X and Z. Regardless of the condition of Y, there can be a closed path from C to A by stipulating the condition of X and Z. For example, if Y is closed $(Y = 1)$, the path can be X, Y, and $\overline{Z}$ where, of course, $\overline{Z}$ is closed. If, however, $\overline{Y}$ is closed, then the path is necessarily X, Y, and Z. In this manner, regardless of the condition of switch Y, it is possible to go from C to A by defining the position of only the two switches X and Z. Since at the moment we are concerned only with the path from C to A, all of the extraneous switches can be removed from the circuit. This results in the circuit shown in Fig. 8-6, which is extracted from Fig. 8-5. An inspection of this circuit shows that it is the same as that shown in Fig. 8-7, with certain sections deleted.

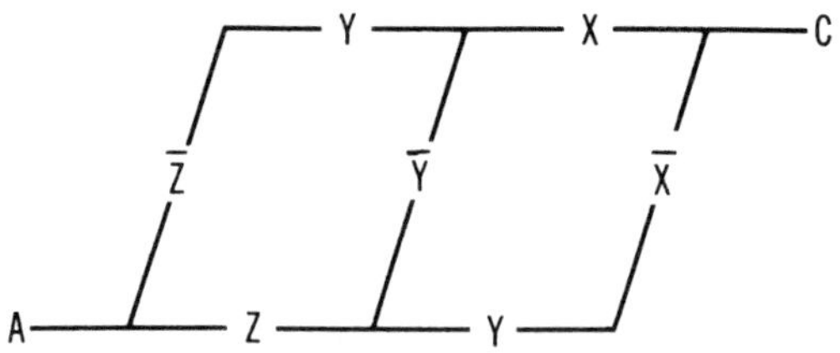

Fig. 8-6. Simplified version of Fig. 8-5.

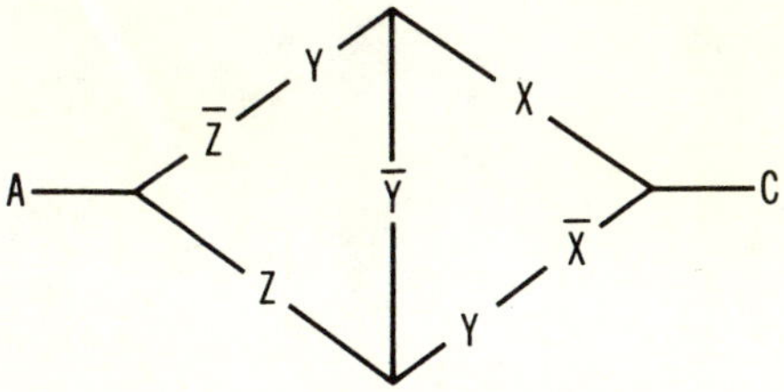

Fig. 8-7. Familiar bridge form of Fig. 8-5.

RELAY CONTROL CIRCUITS

In the preceding discussions, a switch is considered as an idealized circuit element. This means a circuit in which it is possible to open or close a switch instantaneously. In general, most of the discussions in the earlier chapters have assumed a manually operated switch, such as a toggle switch. There are certain limitations, however, of practical switches, and these are extremely significant.

In basic terms an electromagnetic relay, which can be considered a form of switch, has a coil of wire wrapped around a pole piece. There is also an armature that can be attracted or repelled by a magnetic field. The armature is held in certain positions by a spring. In Fig. 8-8, for example, there are three types of relay contacts shown. The contact is the portion through which the current flows, just as in a switch. Fig. 8-8A shows a *make* contact, which is normally open. This means that the spring keeps the armature positioned so that the contacts are not closed during normal operation. When current passes through the relay coil the armature is attracted down, closing the open contacts. Fig. 8-8B shows a break contact, which is normally closed. In this type of arrangement, the spring keeps the armature positioned so that the two contacts provide elec-

(A) Normally open (make).

(B) Normally closed (break).

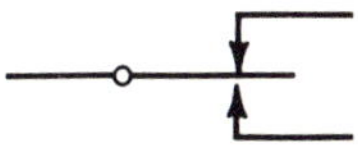

(C) Transfer (break-make).

Fig. 8-8. Types of relay contacts.

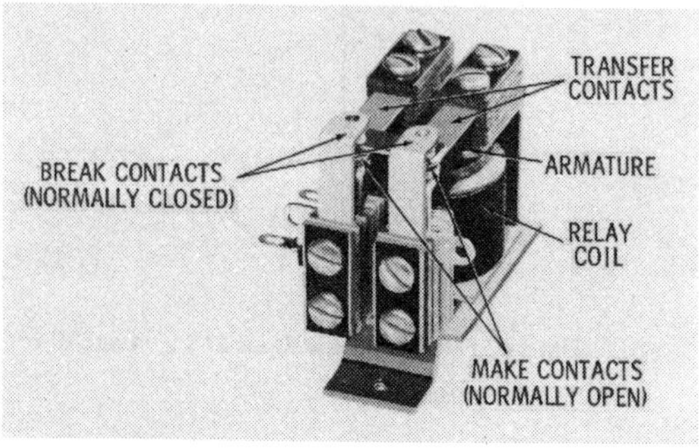

Fig. 8-9. Typical power relay.

trical continuity. When there is current through the electromagnet, the armature is pulled down, opening the contact and breaking the circuit, hence the name break contact.

A combination of these two is shown in Fig. 8-8C; this is a transfer, or operating, contact. In normal operation the armature is arranged so that the upper contact is closed, and the lower contact is open. When current passes through the relay coil, the armature moves to open the upper contact and close the lower contact. Fig. 8-9 shows a typical relay.

A typical relay control circuit is shown in Fig. 8-10. In operation, a switch at A is closed. This causes current from the battery through the relay coil (X) and pulls down both relay armatures. The upper armature closes the contacts marked X. The lower armature closes the lower contacts marked X and at the same time breaks the $\overline{X}$ contacts of the lower set. Thus there are three individual actions that take place in the circuit when switch A is closed.

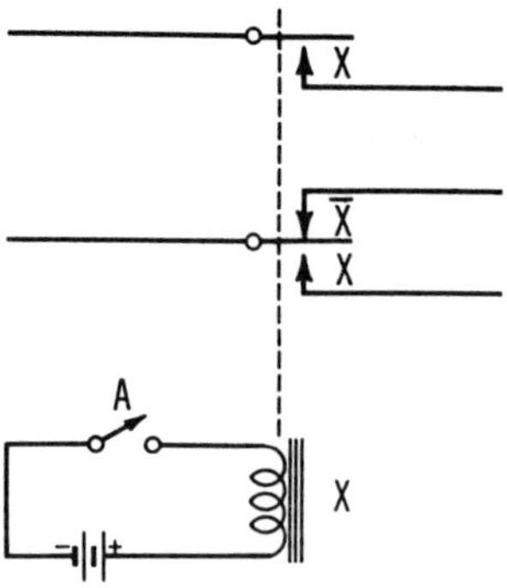

Fig. 8-10. Typical relay control circuit.

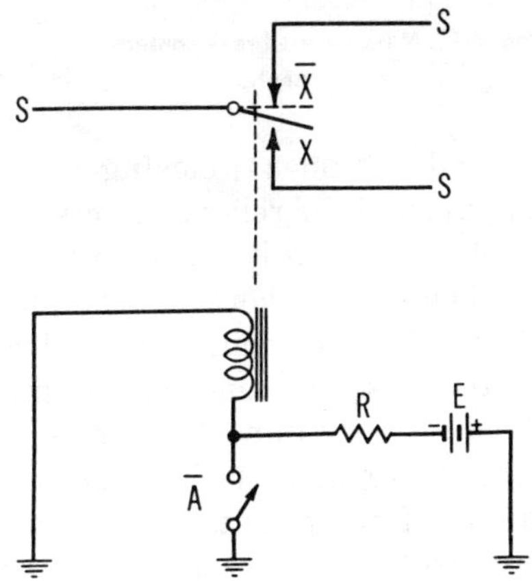

Fig. 8-11. Relay locking circuit.

It is possible to operate a relay on the opening of the switch, as well as on the closing of the switch. Consider Fig. 8-11. There are two parallel paths; one is through the switch, the electromagnetic relay coil, and ground, and the other is through the switch, resistor (R), battery (E), and ground. (The resistor is used to limit the current flow while the switch is closed.) When this switch is open, current from the battery flows through the relay coil and breaks the contacts $\overline{X}$ and engages the contacts marked X, as shown in Fig. 8-11. When the switch is closed, the relay coil is shorted out and the relay contacts return to normal, as shown by the dotted line.

There are a number of relay techniques that are extremely useful in implementing Boolean algebra. One of these is the locking type of circuit shown in Fig. 8-12. Suppose the relay that is normally open has two control switches, one A and the other B. Ignoring switch B for the moment (consider it to be

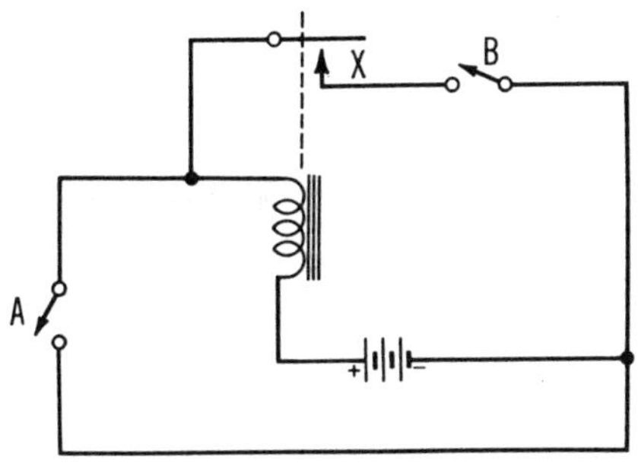

Fig. 8-12. Relay locking-circuit variation.

Fig. 8-13. Make-before-break contact arrangement.

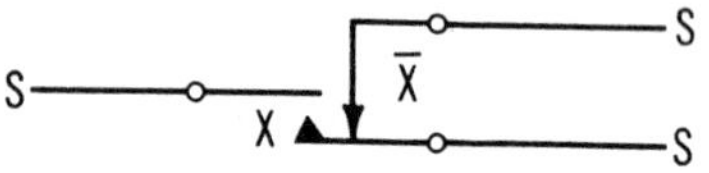

open and left open), closing switch A allows current through the relay, and the relay contacts close. If switch B is now closed, the relay stays in its closed position, regardless of whether or not A is opened. There is a complete path from minus to switch B, through relay contacts X, through the relay coil, through the current source, and back to minus.

Another very important relay-switching consideration is the continuity or make-before-break arrangement. Here there are two sets of relay contacts, X and $\overline{X}$ (Fig. 8-13). When energy is applied to the relay coil, the armature will first move to close the relay contacts X. As the armature continues to move down it will break the relay contacts $\overline{X}$. In this way there is a transfer from one circuit to another through the relay.

Often in relay circuits, a combination of switches can be used to actuate the relay coil. For example, in Fig. 8-14A there are two possibilities; the relay can be actuated by closing switch A, or by closing switches B and C. This is one way of arranging

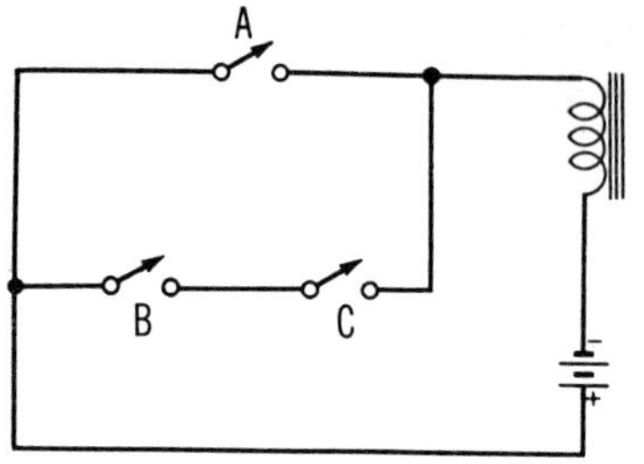

(A) Single relay coil.

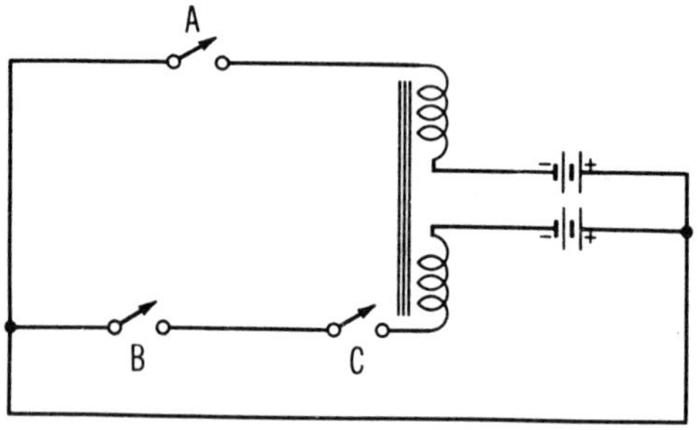

(B) Multiple relay coils.

Fig. 8-14. Multiple switch control of relay.

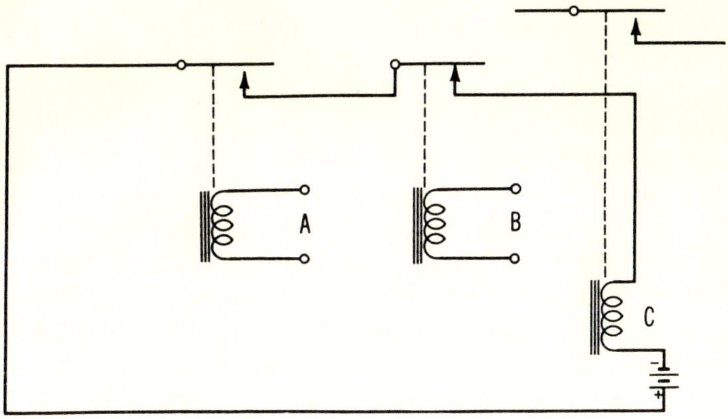

Fig. 8-15. Relay arrangement to obtain $A \cdot \overline{B}$.

the circuit. However, there are cases in which switch A is to be used in several other places, and some form of isolation is needed. This isolation can be accomplished by using a double coil relay, as shown in Fig. 8-14B, with two current sources. This has the same overall operation as the first circuit. However, since A controls its own relay coil and B and C have a second separate relay coil, these two sets are isolated.

A typical relay circuit configuration is shown in Fig. 8-15. There are three relay coils, A, B, and C. As shown, A is normally open, B is normally closed, and C is normally open. There is current through C that actuates its relay contacts if relay

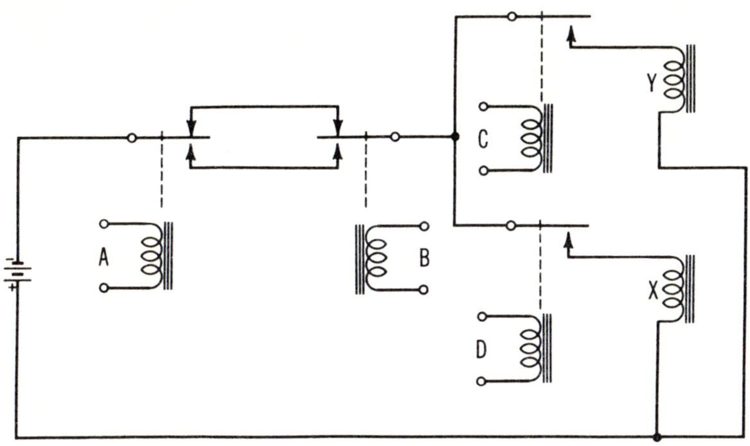

Fig. 8-16. Using relays to implement algebraic expression.

coil B is *not* operated and if relay coil A *is* operated (both conditions occurring at the same time). In this way, by a very simple logical arrangement of A AND NOT B, we can obtain the function C.

Based upon the foregoing information, we can either design a relay circuit from a given algebraic expression or develop the necessary algebra from a given relay circuit. Consider Fig. 8-16; there are two signal relays, X and Y. Each of these controls a set of contacts that perform the work desired from the circuits. These in turn are controlled by relays A, B, C, and D. We will look at this from the standpoint of the circuitry rather than the algebra. Considering only the relays A and B, it is clear from the circuit that there will be no possibility for continuity and current flow unless A and B are in the same state. If the relay coils are both energized, there is current flow; if the relay coils are both de-energized there is current flow. However, if one relay coil is energized and the other is not, there will be no current flow. This is one aspect of the circuit. Now look at relay Y. Relay Y will be operated only if C is energized, and A and B are in the same state. By similar reasoning, X is actuated only if D is energized and A and B are in the same state. This is the same as saying Y is true if C is true and if A and B are both true, or A and B are both false. It is also the same as saying that X is true if D is true and if A and B are both true, or A and B are both false.

CHAPTER 9

The Algebra of States and Events

Previously Boolean algebra has been described in terms of its applications to classes, statements, truth values, and various types of circuit elements. Now we will deal with *states* and *events*, with conditions that change as a function of time. In the real world, this is easily understandable. A cat, for example, becomes a member of the class of cats when it is born; it ceases to be a member of that class when it dies. Similarly, a statement may be true at one time and false at another time. Or a switch may be on in one instant, off in the next.

This new territory is equal to Boolean algebra with the added phenomenon and consideration of time. Quite properly, it could be termed, "Boolean calculus," for like ordinary calculus, it deals with changes in functions. In the earlier discussion of a set of binary variables A, B, and C, we knew that A had to be 0 or 1, that B had to be 0 or 1, and that C had to be 0 or 1. It was evident that A, B, and C each held a single value during the problem—even if that value were not known.

In the algebra of states and events, the same binary variables are used, but they are now considered as functions of time t. As before, A, B, and C must be 0 or 1 during one instant in time. But over an interval of time, the values of A, B, and C may change from 0 to 1, or from 1 to 0. Since the change requires some finite amount of time, the value of A, B, and C during the change is considered indefinite.

UNITS OF TIME

If A, B, and C are considered as functions of time, the first question must be, "How finely is time divided?" Differential calculus suggests that time be treated as a "continuous variable"—that time is to be divided and subdivided into infinitesimal units. While in theory this is also possible in Boolean algebra, it is much more practical to consider time as a succession of finite units. For some applications, it is sufficient to use seconds as the smallest unit of time. For others—especially in the case of the new computer circuitry—it may be necessary to deal in much smaller units of time, such as microseconds, nanoseconds, and probably, eventually, even picoseconds. Both states and events are conditions that last for some duration of time. However, a *state* is usually considered as a condition lasting for a fairly long period of time, while an *event* is a condition, usually lasting only briefly, such as the time required for a circuit element to complete a change of state.

STEP-FUNCTIONS

The binary variables as functions of time can be graphed and are known as *step-functions*, shown in Fig. 9-1. The horizontal axis represents time in whatever units are most appropriate. When the value of the binary variable is 1, a line is drawn; nothing is drawn when its value is 0. For the purpose of our illustration, it is assumed that the horizontal axis represents a span of ten 1-second increments, and that the variable can change its value only at the given intervals.

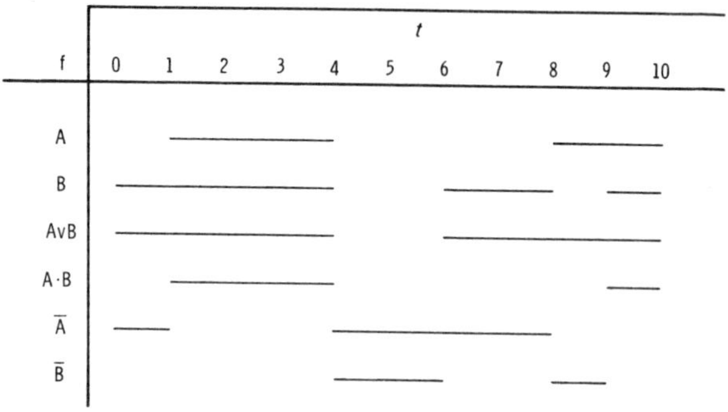

Fig. 9-1. Graphed binary variables or step functions.

Table 9-1. Values for Binary Variables A, B

Continuous Time	Increment t	A	B
0 to 1	1	0	1
1 to 4	2, 3, 4	1	1
4 to 6	5, 6	0	0
6 to 8	7	0	1
8 to 9	9	1	0
9 to 10	10	1	1

The variables assume values during the ten-second time span as shown in Table 9-1. Graphically represented, A, B, and the functions A ∨ B, A • B, $\overline{A}$, and $\overline{B}$, are shown in Fig. 9-1.

At the precise moment a step-function changes, its value is undefined. There is no way of determining whether it is 0 or 1, since its change requires some finite amount of time, small as that may be. In digital circuit design, the practice is not to rely upon the circuit element at that time.

The consideration of time added to Boolean algebra generates a sizeable family of new operators. Some of the more basic members of this new family are:

DELAY	BEFORE	ENTERING	FIRST TIME
HAPPEN	AFTER	LEAVING	SECOND TIME (etc.)

There are no symbols universally accepted for these operators. Common practice is to use the initial capital letter, defining the symbol the first time it is used.

DELAY

The DELAY operator is often used in digital circuitry to account for the small amount of time required for a circuit element to energize or de-energize. Consider statements A and B:

A. Relay X is on at time *t*

B. Relay Y is on at time *t* plus *k*

Statement A has a truth value that varies with time *t*. Associated with relay X is relay Y in statement B. Both relays have the same time pattern (shown in Fig. 9-2), except that statement B is conditional upon statement A and delayed by *k* units

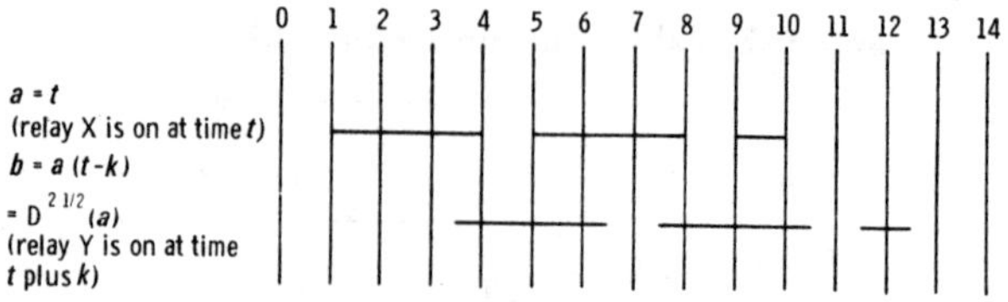

Fig. 9-2. Calculation with the DELAY operator.

of time. Let a represent the truth value of statement A and let b represent the truth value of statement B. Expressed mathematically:

$$b(t+k) = a(t)$$

or,

$$b(t) = a(t-k)$$

For designation of the DELAY operator, it is convenient to assign to it the symbol D.

$$D^k(a) = a(t-k)$$

HAPPEN

Assume that a stands for the truth value of statement A.

A: Switch X is on at time t

Then if H designates the HAPPEN operator, H(a) can be defined as the statement, "The switch X has been on at some time previous to t."

If a and b are two binary variables as graphed in Fig. 9-3,

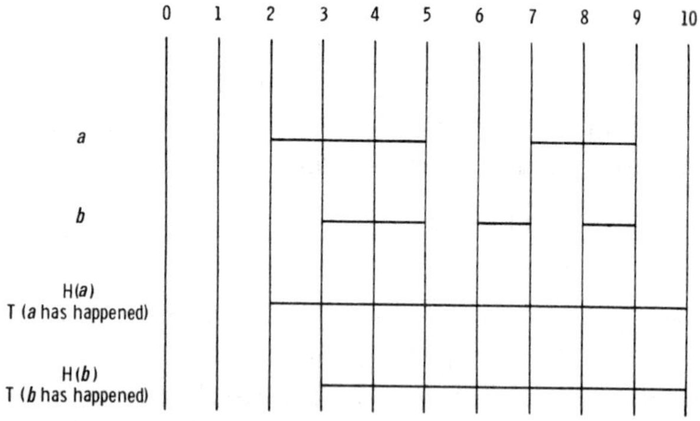

Fig. 9-3. The HAPPEN operator.

then H(a) and H(b) are also binary variables as shown in the same figure.

IDENTITY

The DELAY operator and the HAPPEN operator can be related in an equation with the use of the IDENTITY operator. Let k be a duration of time equal to the time required for a circuit element to safely complete a change of state. Assume that the duration of a is at least as long as k. Then:

$$H(a) = a \vee D^k H(a)$$

If I(a) equals a (I standing for the IDENTITY operator), and k equals 1, then this equation may be written using only operators:

$$H = I \vee DH$$

Another operator results with the negation of the HAPPEN operator, the NOT HAPPENED operator designated NH. It is necessary to distinguish between NH(a) ("a has not happened") and HN(a) ("not-a has happened"). These differences are demonstrated in Fig. 9-4.

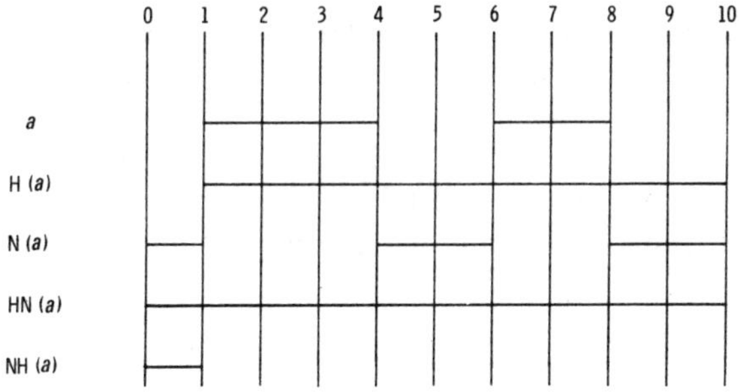

Fig. 9-4. The relation of NOT, HAPPEN, and NOT HAPPENED.

BEFORE, AFTER: ENTERING, LEAVING

In Fig. 9-4, several states and events are displayed for Statement A:

A: Switch X is on at time t, with truth value a

The graph of a, for example, shows that switch X is on for times 1 to 4, 6 to 8, and off for all other times within our ten-unit frame of reference. There are five states: three off and two on. There are four events that occur at times 1, 4, 6, and 8.

The operator BEFORE (initial B) is used to indicate that "a has not yet happened." Thus:

$$B(a) = NH(a)$$

The operator AFTER is used to indicate "not a, yet a has happened previously." Using A to represent AFTER:

$$A(a) = H(a) \cdot N(a)$$

Switch X *enters* the state of being off at time 0, and *enters* the state of being on at times 1 and 6. To define the use of the operator ENTERING, let k represent the time required by a circuit element to safely complete changing its state. Thus a definition of E(a) where E represents the operator ENTERING is "a, but not yet D^k (a)," or:

$$E(a) = a \cdot ND^k(a)$$

Similarly, a definition of the operator LEAVING, if L represents that operator, is "not a, but still $D^k(a)$," or:

$$L(a) = N(a) \cdot D^k(a)$$

These last two operators are very useful in certain ambiguous cases where a would necessarily be indefinite. As k gets smaller and approaches zero, both E(a) and L(a) can still have the truth value of 1. However, if k were to reach zero, $a = \bar{a}$—which in Boolean algebra equals 0. This never happens in dealing with circuit elements, since some small amount of finite

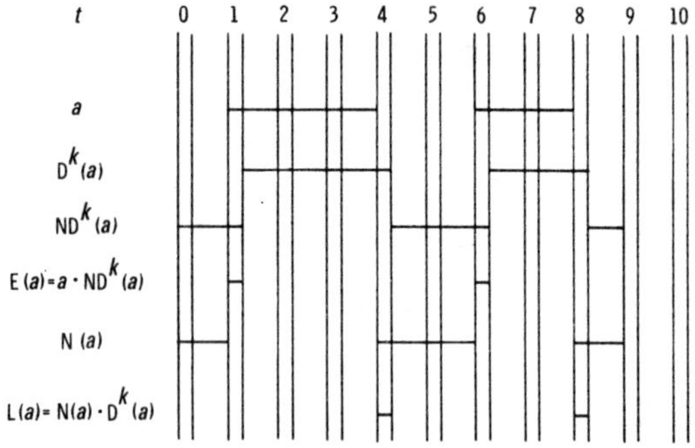

Fig. 9-5. ENTERING and LEAVING.

time is always required for one circuit element to influence another. Fig. 9-5 illustrates the operators ENTERING and LEAVING, where k is assumed to be about 1/6 of the interval used to measure time, t.

FOR THE FIRST TIME, FOR THE SECOND TIME

If a binary variable has several states or stretches of being on, it may be necessary to distinguish between these several stretches. Let I^z represent FOR THE Z TIME:

$I^1(a)$ represents, "a for the first time"
$I^2(a)$ represents, "a for the second time"

and so forth.

In this same system of notation,

$H^2(a)$ represents "a has happened for the second time."

These operators can then be defined in terms of the operators DELAY, HAPPEN, and NOT HAPPENED previously defined. $I^1(a)$ is calculated and displayed graphically in Fig. 9-5. Assume that $X(a)$ is 1 if $a_0 = 1$, and that $X(a)$ is 0 if $a_0 = 0$. Both of these cases can be represented in a single formula:

$$I^1(a) = [X \bullet I \bullet NH(N \bullet H) \vee NX \bullet NHN]\,(a)$$

Of course it is true that

$$I^1(a) + I^2(a) + I^3(a)\ \ldots = I(a)$$

as all of the stretches of a taken together must equal a.

The calculation of $H^2(a)$ in the instance where $a_0 = 0$ is shown in Fig. 9-6, and expressed in the formula:

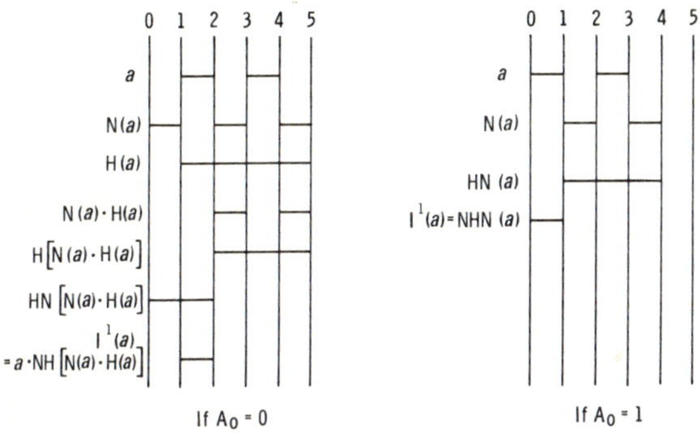

Fig. 9-6. Calculation of I¹ (a).

$$H^2(a) = H[I \bullet H(N \bullet H)](a)$$

The operators discussed in this chapter could obviously be combined to produce many more, but these illustrate the laws of Boolean algebra governing states and events. It should be remembered that in performing Boolean operations on states and events, outputs may be functions of logic that existed sometime previously. Circuits such as these are often termed "sequential systems," and use what is termed "sequential logic."

In sequential logic, time is quantized into clock pulses or clock times. A possible sequential expression might be:

$$F_{t=5} = (A_{t=1} + B_{t=3}) \bullet (C_{t=2} + F_{t=4})$$

This expression implies that F at time interval 5 is a function of A at interval 1, B at interval 3, C at interval 2, and F at interval 4. From this expression, it is evident that some form of memory, or storage, is required. In the expression above, the function F at interval 4 must be stored until interval 5. A generalized memory element such as a flip-flop can be defined as having the following inputs and outputs.

Set to 1 (S)	1 OUTPUT (F)
Reset to 0 (R)	0 OUTPUT (F)

The Set and Reset inputs are activated only one at a time, and only one output is active at a time. The flip-flop, or bistable multivibrator, remains in its previous condition until the opposite condition is presented to its input. In a truth table (Table 9-2) for these conditions the notation F signifies the previous state or time period. Consider the equation for F derived from this table:

Table 9-2. Truth Table for a Sequential Expression

S	R	$\underline{F}$	F
0	0	0	0
0	0	1	1
0	1	0	0
0	1	1	0
1	0	0	1
1	0	1	1

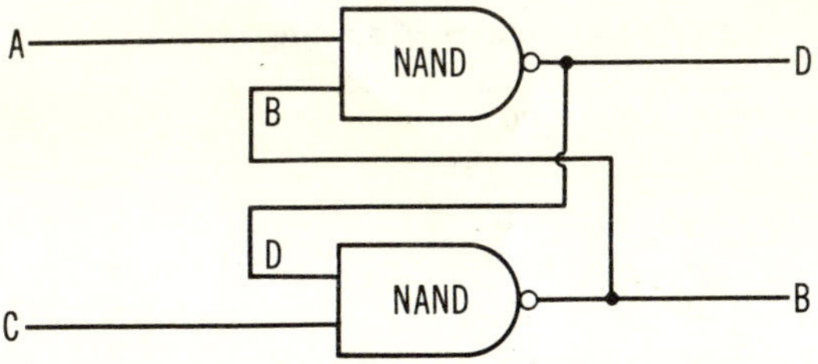

Fig. 9-7. A flip-flop constructed from NAND circuits.

$$F = \underline{F} \cdot \overline{S} \cdot \overline{R} + \underline{\overline{F}} \cdot S \cdot \overline{R} + \underline{F} \cdot S \cdot \overline{R}$$

This reduces to:

$$F = \overline{R} \cdot (\underline{F} \cdot S + S)$$

NAND/NOR circuits can be used to synthesize a circuit whose equation is that of a memory flip-flop. Utilizing a cross-connected pair of NAND circuits, a flip-flop can be produced as in Fig. 9-7. The equations for the circuit are:

$$D = \overline{A \cdot B}$$

but,

$$B = \overline{C \cdot D}$$

hence

$$D = \overline{A \cdot (\overline{C \cdot D})}$$

In this flip-flop consider these relations:

1. D will be active (logic 1) if A is logic 0, or if C is logic 1 and D was logic 1.
2. D will be inactive (logic 0) if A is logic 1 and C is logic 0, or D was logic 0.
3. The set condition occurs if the flip-flop was set and no reset occurs, or if a set occurs as a logic 0 level at the A input.
4. The reset condition occurs if the flip-flop was reset and no set occurs, or if a reset occurs as a logic 0 at the C input.

This flip-flop is the basic SR FF (Set-Reset Flip-Flop), but many other configurations can be derived and employed. One common alternative is the counter flip-flop. This device contains, in addition to, or in place of, the S and R inputs, a trigger

Table 9-3. Binary Counting Sequence

2^2	2^1	2^0	Dec
0	0	0	0
0	0	1	1
0	1	0	2
0	1	1	3
1	0	0	4
1	0	1	5
1	1	0	6
1	1	1	7
0	0	0	0

input. The trigger input is an ac input requiring a pulse going from the logic 0 to the logic 1 state to cause activation. The trigger action causes a change of state of the flip-flop, *regardless of the previous condition.* With this action they are sometimes called complementary flip-flops.

The trigger flip-flop is used in counting circuits. The binary counting sequence is shown in Table 9-3 with the equivalent decimal values. Inspection of this table shows that a next higher order bit changes state when the lower bit changes from a 1 to a 0. The flip-flop counter for 3 bits is shown in Fig. 9-8. The trigger input was defined to require a logic 1 activation. Interstage connection is between the T (trigger) input and the 0 output of the preceding stage. The 0 output is the logic inverse of the 1 output; when the 1 output is energized, the 0 output is not. Thus, a transition of the flip-flop from the 1 to the 0 state causes the output to go from the 0 to the 1 state; thus it meets the requirements for a trigger.

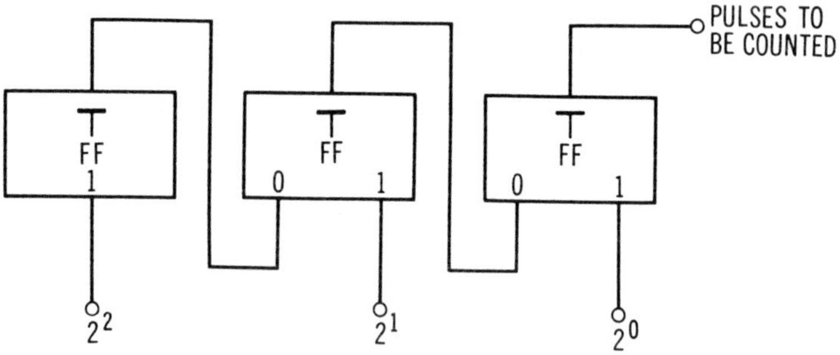

Fig. 9-8. A three-bit, flip-flop counter.

Index